D1093349

Limulus in the Limelight

A Species 350 Million Years
in the Making and in Peril?

Limulus in the Limelight

A Species 350 Million Years in the Making and in Peril?

Edited by

John T. Tanacredi
National Park Service
Brooklyn, New York

Introduction by

Sylvia A. Earle

and Conclusion by

Niles Eldredge

Kluwer Academic / Plenum Publishers
New York, Boston, Dordrecht, London, Moscow

Library of Congress Cataloging-in-Publication Data

Limulus in the limelight : a species 350 million years in the making and in peril? / edited
by John T. Tanacredi ; introduction by Sylvia A. Earle ; conclusion by Niles Eldredge.
    p. cm.
    Includes bibliographical references (p.).
    ISBN 0-306-46681-3
    1. Limulus polyphemus. 2. Endangered species. I. Tanacredi, John T.

QL447.7 .L56 2001
595.4'92--dc21

                        2001038768

ISBN 0-306-46681-3

©2001 Kluwer Academic / Plenum Publishers, New York
233 Spring Street, New York, New York 10013

http://www.wkap.nl

10 9 8 7 6 5 4 3 2 1

A C.I.P. record for this book is available from the Library of Congress

Printed in the United States of America

This book is dedicated to John Loret, Ph.D.,
scientist, teacher, seaman, explorer, educator and dear friend.

Foreword

MARC KOENINGS

General Superintendent, National Park Service, Gateway National Recreation Area

The type and scope of cooperators responsible for this publication is indicative of the effort we as environmental stewards must take to protect the fascinating horseshoe crab. The National Park Service is redoubling its efforts to "preserve and protect" our Nations' natural resource base, especially if history is to be our guide and we take a lesson from the Carrier Pigeon, California Condor and Right Whales; The Explorers Club whose conservation message is borne out through scientific studies at every corner of the globe, through the wonder and excitement generated by exploring our world, (to see children looking at horseshoe crabs along the shoreline and you'll know what I mean); to The Science Museum of Long Island whose logo "It's All About the Children" brings home to all of us the critical message of "conservation begins with education" and beginning as early as possible to learn about "life and the universe" will be the critical factor in shaping our future conservation ethic; to the VIMS Horseshoe Crab Museum and Repository where the scientific minds can connect to the "computer mind" and search for those scientific gaps we need to fill in about our knowledge of this keystone species; and, lastly those conservation organizations which are on the "frontline" unencumbered with the bureaucratic rumblings that sometimes can cloud a critical issue into an inactive state – it's the groups such as New York City Audubon, the American Littoral Society with grassroots concern and monitoring of horseshoe crabs which prompted the National Park Service at Gateway National Recreation Area into action over five-years ago to set up mini-conferences to investigate the horseshoe crab population problems, to the educational and societal groups like AREAC (Aquatic Research and Environmental Assessment Center) at the CUNY Brooklyn College Campus or MACUB (Metropolitan Area College and University Biologists) who

vii

shared our concerns and helped co-sponsor a mini-conference at the NYZS Wildlife Conservation Societies' Aquarium for Wildlife Conservation (part of the Coastal America program).

The conservation community makes significant efforts each year to protect threatened and endangered species against the onslaught of a burgeoning global population, excessive resource consumption and a myriad of pollutants. To be anthropocentric for a moment, when a species that has dramatic impact on human lives (i.e. research on vision and bacterial contamination) this alone should prompt us to make extraordinary efforts to protect and preserve this species for future generations. Having worked in the coastal zone for over 30 years I believe what is proposed here for the protection of *Lumulus* goes beyond the idea of its continued existence, (which frankly could be enough to justify protection status). It goes to our philosophy about earth's natural resources and life in general. Common sense alone should dictate our maximum efforts to prevent this organism from going over the brink into extinction. I applaud the presentations in this book and I will recommit our resources toward preserving these incredible creatures into the new Millennium.

Acknowledgments

Special thanks go to Dr. John Loret, Director, The Science Museum of Long Island, and Kevin Buckley (retired) past General Superintendent, Gateway NRA for their support of the three mini-conferences resulting in this book.

For providing external review of the first drafts of this work – P.A. Buckley, Betty Borowsky, Robert Cook, Chris Boyko, Mark Matsil, Raul Cardenas, Mike Soukup, Chikashi Sato, George Frame, Martin Schreibman - my sincerest thank you goes out to your contributions and efforts.

A special thank you for our U.S. Park Police Marine Unit, Gateway NRA and SCUBA divers for providing video work done in Jamaica Bay – Chris Pappas, Tom Pellinger, Grant Arthur, Paul Dorogoff and the Division of Natural Resource Fisheries Biologist, Christine Kurtzke. Photographs were provided by Don Riepe, Christine Kurtzke, Carl Shuster and myself.

A special thank you for the manuscript preparation to Joanie Crane, whose patience is only surpassed by her professionalism and perseverance in getting this book to completion.

A thank you to MACUB – Metropolitan Area College and University biologists, AREAC – Aquatic Research and Environmental Assessment Center, Brooklyn College, CUNY, NYC Audubon, American Littoral Society, New York Aquarium for supporting the three mini-conferences culminating in this publication. Please be reminded that all errors of content or omission are mine alone.

Finally to Joanna Lawrence at Kluwer Academic/Plenum Publishers for her guidance, foresight and professional leadership in getting this book to its completion. It could not have been done without her.

John T. Tanacredi, Ph.D.
Editor

Horseshoe Crabs congregate in their largest numbers along the New Jersey shoreline in Delaware Bay. Photo by Don Riepe.

The crab in the center of this photograph, encrusted with barnacles, has completed its molts. Photo by Don Riepe.

Males mob a female in the intertidal marsh. Photo by Christine Kurtzke.

Laughing Gulls feeding on Horseshoe Crab eggs. Photo by Don Riepe.

Contents

PART IV: PRESENT DAY INVESTIGATIONS

PART V: A FINAL WORD

Introduction: *Limulus* in the Limelight

SYLVIA A. EARLE
Deep Ocean Exploration and Research, Oakland, California

Late in May 1996, I was lured to the shore of New York at Jamaica Bay by the promise of seeing some old friends – horseshoe crabs by the hundreds – and John Tanacredi, Research Ecologist for the National Park Service at Gateway NRA in New York/New Jersey. By horseshoe crab standards, John is by no means "old" but we've known each other for years, long enough for me to have learned that when John says,

"You've got to come" . . . he had good reasons. So, early in the morning on a rising tide, John and I joined a group of other crab-watchers and started down a grassy path leading to the beach where waves of shorebirds, a scattering of cormorants, ducks, and geese, several dozen glossy ibis, and hundreds of gulls and terns were already up and about, feasting on the newly-deposited wealth of horseshoe crab eggs.

Packed within each jade-like sphere, gleaming like miniature translucent peas, the eggs provide not only the vital ingredients for the next generation of crabs, but also sustenance for millions of birds. Some have become dependent on the eggs as a critical mid-migration source of energy during flights from the tip of South America to nesting sites in the Arctic. Together, the birds and crabs make up one of the longest (more than 10,000 miles), slenderest (the intertidal area and migration pathway), and most ephemeral (a few weeks each year) ecosystems on earth.

For me, the visit to the beach was a personally historic journey, a reminder of the sense of wonder I experienced encountering horseshoe crabs during family vacations to the New Jersey shore when I was a child. I was enchanted by their craggy, exotic beauty, their curious eyes, sturdy sword-like tail, the encrusted cargo of slipper shells, barnacles and seaweed, but most of all, their stoic attitude when interrupted from whatever they were

Limulus in the Limelight, Edited by John T. Tanacredi
Kluwer Academic/Plenum Publishers, New York, 2001

doing to endure inspection by a young, curious primate, and swift return to business as usual as soon as I let them go. It pleased me to be able to instruct frightened adults about the crabs' benign nature, and to sometimes enlist their help in rescuing stranded or overturned individuals.

As if seeing my own past come alive, ahead of us along the shore, a young girl bent down and wrapped her small fingers around what appeared to be half a basketball, then lifted from the wet sand a wondrous creature glistening like polished mahogany, *Limulus polyphemus*, an Atlantic horseshoe crab, a descendent of a line of arthropods that have changed little in 350 million years. Before the Americas split from Eurasia and Africa and formed the Atlantic Ocean, before dinosaurs, birds and mammals, before flowering plants, before even the faintest whisper of the future of humankind, there were horseshoe crabs. They survived waves of extinction including the catastrophic changes 65 million years ago that marked the end of the dinosaurs and many other forms of life, yet now face perils from people – pollution, habitat destruction, incidental capture in fishing nets and deliberate taking by the ton for use as fertilizer and bait -- that could soon cause their demise. The enormous number of ten to twenty year old adults intentionally or accidentally killed in recent times threatens not only the future of *Limulus,* but the entire suite of creatures – birds, fish and invertebrates – that they help sustain.

Fossils give clues about the existence of numerous kinds of horseshoe crabs, all with the same distinctive shape, and most apparently aquatic, although some, now long gone, may have ventured for a time onto the land. Not crabs at all, but rather, related to the ancestors of spiders and scorpions, horseshoe crabs today pack no venom nor are their legs tipped with effective claws. Only four species remain as living evidence of a rich and distinctive heritage of organisms defined as the Class Merostomata, a group that is given the same genetic heft by scientists as the Class Insecta with some 500,000 species. Loss of one insect species means losing 1/500,000 of the diversity of insects; loss of a species of horseshoe crabs translates to losing a quarter of that entire branch of life.

Three Asian species are rarely seen today and only a few thousand individuals, all in all, are believed to exist in the Indo Pacific region of the world. But what of the single Atlantic species – so numerous at the beginning of the 20th century that millions were collected for fertilizer and millions more crushed for road building materials? This volume puts *Limulus* in the limelight, bringing together the work of various experts concerned about the robust past and recent swift decline of these durable – but vulnerable – creatures, as well as recommendations about what to do to ensure the continued survival of their ancient line of life. Several authors address the value of the unique attributes of *Limulus* to humankind for medical analysis and products, while others focus on their behavior and natural history and their value in education. The collection together represents not only a hearty salute to horseshoe crabs of the past and present, but also is a haunting reminder of the power we have over their future.

If in this geological moment we allow the loss of creatures that represent the distillation of hundreds of millions of years of fine tuning, it suggests something terribly wrong about human priorities. And it suggests something wonderfully right if, through our conscious caring, they – and we—are given a chance to continue to share space on Earth for millions of years to come.

PART I

Local Species with Global Implications

Photo by Don Riepe.

"Is it possible that humanity will love life enough to save it?"
in *Biophilia*, E.O. Wilson (1984: 185).

Chapter 1

Horseshoe Crabs Imperiled?
The Fate of a Species 350 Million Years in the Making

JOHN T. TANACREDI
National Park Service, Gateway National Recreation Area, Division Natural Resources, 210 New York Avenue, Staten Island, New York 10305, USA

1.   INTRODUCTION

In an era when sports numbers shatter historical ceilings (especially in baseball) and fall by the wayside like fall leaves, a most enduring record of life continues to crawl along emerging from the sea early each spring pronouncing their incredible "record" of existence...over 350 million years. Now there is a number that even the rarified air of baseball salaries can admire. Yet, today the remarkable sea creature which endured over 100 million years of dramatically changing earth history prior to the existence of the dinosaurs, and which is comprised of only four species globally, is being tested for its' survivability as at no other time in their known history. Human activities pose the most significant threat to *Limulus polyphemus*, the only North American horseshoe crab species, and have dramatically impacted the other three species found along the southeastern shore of Asia; so much so that in Japan for example only a few thousand of a single horseshoe crabs species remain.

The biology and "uses" of horseshoe crabs have been written about for sometime. I remember the stories explained to me of their biological "oddities" by docents at the New York Aquarium (part of the then New York Zoological Society, today entitled the Wildlife Conservation Society). As a born and bred Brooklynite, I always felt the Aquarium in Coney Island was kindred and should have been christened the name of its location. (I still haven't gotten over the Brooklyn Dodgers leaving Brooklyn, so bear with me...) Except for possibly research covering their "compound eyes" which

Limulus in the Limelight, Edited by John T. Tanacredi
Kluwer Academic/Plenum Publishers, New York, 2001

were the subject of the 1967 Nobel Prize in science exploring the human eyes' mechanics, to the telson (tail) used to right the overturned crab and not "shoot their stinger at you" when you wade into the shoreline shallows, there still is very little that is known about these unique organisms' ecology, except possibly in its mutualistic relationship to migrating bird populations. In fact, the most dramatic recent-time restriction on horseshoe crab adult harvesting was prompted in New Jersey by the reduced shore bird numbers observed along the southern New Jersey shoreline for over five years. New Jersey Governor Whitman temporarily closed down the eel fisheries in Delaware Bay and along the New Jersey shore so that an assessment of the horseshoe crab populations could be conducted. The restriction didn't last long, and in 1999, 2.5 million horseshoe crabs were "legally" (by permit), collected for New Jersey eel potters use. (Niles, L., 2000)

Some of the most basic biological information about a species still eludes us in describing horseshoe crabs. The aging of horseshoe crabs, for example, has in many respects been an indirect method. It takes approximately 10 years for horseshoe crabs to mature to adulthood, molting off youthful carapaces 16 to 17 times depending upon its gender; females molt seventeen times while males molt sixteen times. The significance of this is still unknown and could be more than just polymorphism. Once molted to adulthood they will not molt again for the remainder of their approximately 20-year lives. The age of a horseshoe crab is usually determined by the level of encrusting organisms (i.e., bryozoans, slipper shells, algae, etc.), by the number of species and concentrations of these species on their shells. Later on in this book you will be able to read about the hanger's-on to the horseshoe crab shell.

Limulus' linkage to the shore blends into the biology of a number of species. The life cycle of this invertebrate affects the life cycle of, and is synchronized with the life cycle of several vertebrate species; not the least of which are humans. The first time I observed the "shorebird season" in Delaware Bay on a short beach near Cape May New Jersey, it was truly a shorebird convention with species like Red Knots (*Calidris canutus*) on their way to Canadian arctic breeding grounds coming from as far as the tip of South America. The shoreline appeared to be frothing as if a clothes washer went haywire and discharged soapy waters and bubbling detergents along the coast. This was due to the churning of the waves and their breaking right upon the horseshoe crabs as they release eggs and sperm.

Since 1990 the spawning population in Delaware Bay, which has a history of harvesting crabs dating back to the 1870's when over 4 million horseshoe crabs were taken from the Bay, has declined. Both conch (*Strombus gigas*) and eel (*Anguilla rostrata*) fisheries continue to harvest females rather than alternative baits and will harvest immature horseshoe crabs, which molt for the last time in the fall. Efforts to allow harvesting offshore primarily after spawning season is counter productive in protecting this species since harvesting prior to their first spawning season eliminates the most potentially fertile portion of the spawning population. (Berkson, J. and C. Shuster, 1999)

2.   AVIAN CONNECTIONS

Due to the dependence of coastal migratory shorebirds such as Red Knots (*Calidris canutus*), Ruddy Turnstones (*Arenaria interpres*), Sanderlings (*Calidris alba*) and Semipalmated Sandpipers (*Calidris pusilla*) on the food resource provided by millions of horseshoe crab eggs laid at the critical time of peak shorebird migration, these eggs of horseshoe crabs become the primary energy source (weight gains of 40% reported), (Castro, G. *et al.*, 1989) to support their marathon migration north to the Arctic.   The Horseshoe crab densities of Delaware Bay have not been found in any other locations on the Atlantic Coast from Maine to Florida.   Surface horseshoe crab egg densities is the prime factor in attracting up to 1 million shorebirds each year since 1980 when records started being taken for Delaware Bay (Clark, K.E. *et al.*, 1993). The significance of the shorebird impacts globally cannot be overstated.   There are approximately 50 shorebird species out of a known total of 177 species in the Western Hemisphere and they fill distinct and unique coastal shoreline niches.   Mud birds or open tidal flat feeding birds (i.e. Black-bellied Plover); wading birds in shallows of wetlands (i.e. Avocets, Yellowlegs and Phalaropes) and the coastal birds which include Sanderlings and Oystercatchers; surfbirds like the Piping Plover; upland birds such as Killdeers, (Corven, J., 1998). Shorebirds generally follow specific flight paths during their migrations north, stopping at areas for feeding purposes.   Birds that winter in Argentina and Brazil along the Atlantic Coastline will follow the Atlantic Flyway onto the Arctic.   Starting in Argentina, upward of 30% of the flyway population, transit Brazil and then on to Delaware Bay to obtain 30% body fat from copious quantities of Horseshoe crab eggs, and then, on to Hudson's Bay or further north to meet up with the summer hatching of flies in their nesting sites in Canada. These stopovers support more than 500,000 shorebirds each year, (Corvin, J., 1998).

3.   ATLANTIC COAST EXPERIENCES

Colonial American fishermen, who used the carapaces to bail out their boats, called them pan crabs. *Limulus polyphemus* is the only "American" species found from Nova Scotia to Florida and into the Gulf of Mexico. Three other species live along the shores of Asia and the East Indies. Sixty million years ago all horseshoe crabs were concentrated in European waters. Then, for some unknown reason or reasons, they migrated never to return. Thus the geographic distribution of all four species in only 2 regions. North American Indians often attached the spike-tail to the ends of their fishing spears as spearheads.

Studies of the Horseshoe crab's eye have also given scientists vital information about eye disorders such as retinitis pigmentosa, which causes tunnel vision and can lead to total blindness. It is the study of Horseshoe crab's blue blood caused by a pigment called hemocyanin which causes the blood to turn bluish color upon exposure to air. Limulus lysate a blood fraction produced only by the horseshoe crab is used in research on spinal meningitis, blood clotting and cancer. Delaware Bay, the epicenter of horseshoe crab-dom is subject to an annual census of horseshoe crabs. The Delaware Bay population estimated at 1.2 million in 1990 had dwindled to less than 200,000 by 1995, (Horton, T., 1996).

Horseshoe crabs are the ultimate generalist, more than likely their secret to their longevity revealed. However, considerable questions remain to scientists who have spent their lives studying these creatures. They are euryhaline, can tolerate a wide range of temperatures, can live on just about any type of prey, and yet are only found on the eastern shore of the United States and shorelines of Asia and Japan. Why aren't they found on every shoreline? Today only four recognized species of horseshoe crabs are still living; the fossil record for horseshoe crabs is not diverse and there does not ever seem to be more than a handful of species on earth at any one time. How is success determined in the history of life on earth? By the number of species produced? Or by the longevity of a group? We still have a lot to learn from these resilient species.

4.   RESEARCH

Captured crabs are taken to laboratories where blood samples are taken from approximately 250,000 crabs annually for the biotechnical industry worth an estimated $50 million annually. The females lay their eggs (upward of 4,000 per female) up to 8" deep in the sand deposited at the level of the beach reached only by the high tides which occur during the new and full moon. Laying eggs at the high water mark protects the eggs from fish predation such as eels and minnows and juvenile Striped Bass, but it is the 95,000 Red Knots estimated to gather at Delaware Bay consuming an estimated 248 tons of horseshoe crabs that places predation pressure on horseshoe crabs.

Increases in horseshoe crab catches have exponentially outpaced any state restrictions on total numbers allowed to be "harvested". In Virginia alone the 1998 catch was 28 times the historical catch, from 25,000 to over 750,000 crabs. This pressure has by all accounts reduced the observable horseshoe crab population by 50% since 1990. Though this is mostly anecdotal from scientists and regulators, it is a universal response from coastal researchers and people who have observed horseshoe crabs for decades as teaching tools; for example the interpretive Park Rangers at Gateway NRA, Fire Island National Seashore, Cape Cod and Assateague have indicated that they "haven't seen as many horseshoe crabs as there used to be". This unfortunately is pretty much the best we have at this time. Biologists still don't know how often female horseshoe crabs lay eggs each season, the length of the females reproductive life, how storms, dredging or other beach disruptive processes affect nesting beaches and reproduction, or even how many crabs come into Delaware Bay or New York Harbor. If we are to have a sustainable population of horseshoe crabs, these and other questions must be answered.

Previous Department of the Interior Secretary Bruce Babbitt visited Pickering Beach on the Delaware in May, 1999 and wrote a letter to the Atlantic States Marine Fisheries Commission urging them to consider strict regional limits on crab kills to all Atlantic States, (Ward, P.D., 1991). Besides over harvesting, shoreline dynamics and human development have reduced horseshoe crab habitat. Some scientists have indicated that sea level rise due to global warming phenomena may also be impacting on horseshoe crab numbers. Sandy beaches are also dwindling due to erosion and general shoreline retreat. Without reversing these trends, over-harvesting will potentially prove the deathnell of a species that has survived for thousands of millennia.

Scientific research, in support of future management of horseshoe crab population, is sorely needed. Due to the limited "season" for horseshoe crabs, observations are varied in determinations of what would appear to be a fairly simple piece of data for these organisms; namely, the average sex ratios for breeding populations. Different values have been reported for spawning beaches, (Harrington, B. and C. Shuster, 1985) of 5:1 and 3:1 (male to female) to 1:1 for an overall sex ratio estimate, (Shuster, C. and M. Botton, 1985).

There are few regulations protecting this animal along the east coast of the U.S. Based upon reported landings (which are notoriously underestimates of actual removal) in New Jersey alone, (Rudloe, A., 1980) horseshoe crabs harvests have increased in three years from approximately 250,000 in 1993 to over 600,000 in 1996.

John T. Tanacredi

Table 2. Atlantic states landings for horseshoe crab for the period 1970-1997.

ATLANTIC STATES LANDINGS (MAINE-FLORIDA)

Year	Pounds	Value (in $1000s)
1970	15,900	7.79
1971	11,900	3.01
1972	42,000	2.63
1974	88,700	5.54
1975	6,700	6.90
1976	62,800	18.90
1977	2,043,100	63.96
1978	473,000	16.58
1979	728,500	45.59
1980	1,215,630	148.24
1981	566,447	79.02
1982	326,695	55.97
1983	510,060	44.95
1984	440,959	35.83
1985	152,392	15.36
1986	522,199	41.46
1987	507,814	47.82
1988	462,663	67.82
1989	636,252	71.23
1990	1,087,912	131.72
1991	908,130	101.81
1992	1,089,045	121.50
1993	1,000,619	109.71
1994	1,906,059	207.22
1995	1,401,656	228.60
1996	2,547,987	378.99
1997	1,885,883	334.44

Source: National Maine Fisheries Service Plan (1998)
Note: All dollars are 1992 dollars, adjusted by the implicit price deflator (GDP). All life stages are included.

In 1989, the FDA reported that 130,000 horseshoe crabs were used in the biomedical industry. The current estimate of medial use is around 250,000 of which about 10 % of the crabs do not survive the bleeding procedure,

(ASMF Commission, 1998). All crabs are to be returned to the environment, however not necessarily where they were collected.

Horseshoe crabs have been used since the 1950's for chitin coatings of suture materials which have been shown to enhance healing by 35-50%, (Rudloe, A., 1980; NMFP, 1998). There is limited harvesting for blood in leukemia and other cancer research as well as other research associated with endotoxins and other control therapies.

It has been fairly well established that in the Delaware Bay estuary between mid-May and early-June intertidal beaches are fairly well filled with migratory shorebirds whose most abundant food items is horseshoe crab eggs, (Botton, M. *et al.*, 1994). The emergence of trilobite larvae of horseshoe crabs *Limulus polyphemus* is usually correlated with full moon or unusually heavy wave action, (Rudloe, A., 1978). A small spawning population persists through August whose developmental rate of embryos is temperature dependent, (Jegla, R.C. and Costlow, J.W., 1992). Investigations have shown that many trilobite larvae will survive over winter in sands when bird predation is minimal, (Botton, M. *et al.*, 1992). Sediment disruptions could influence survivorship of these over-wintering embryos.

Horseshoe crab habitat is from "muddy" (silt-clay) bottoms (when not spawning) to a gravely/shell substrate. When a sandy beach is not available for spawning, the female will attempt to make nests in a beach of oyster shells or gravel, but will tend to avoid areas of exposed peat. In Chesapeake Bay, the juvenile loggerhead turtles that summer in the bay feed heavily on horseshoe crabs. Small fish also feed extensively on the eggs and larvae of *Limulus*. Other predators include such disparate creatures as sharks, sea gulls, boat-tailed grackles and raccoons. Horseshoe crabs are therefore a significant component of the coastal ecosystem of our North American Atlantic estuaries, (Prior, R. B., 1990).

5.   A FINAL WORD

All in all, we have the classic symptoms of anthropocentric perceived needs driving a species down a track to extinction. Some may say this is an overstatement and that we have enough time to develop a management plan for this species. Our track record unfortunately in protection or preservation of sustainable populations in natural systems does not reflect this as a salient approach. There must be a moratorium on horseshoe crab harvesting for a minimum of 3-5 years coupled with an intensive investigation into an appropriate management scheme. Only then can we make a more informed decision on the protection and long-term preservation of a species that has taken 350 million years to make only to be undone in less than a century.

REFERENCES

Berkson, J. and C.N. Shuster (1999) "The Horseshoe Crab: the battle for true multiple-use resource" Fisheries 24 (11): 6-10

Botton, M.L., R. Loveland and T.R. Jacobsen (1994) "Site Selection by Migratory Shorebirds in Delaware Bay, and its Relationship to Beach Characteristics and Abundance of Horseshoe Crabs (Limulus polyphemus) Eggs" The Auk; (3): 605-616

Botton, M.L., R.E. Loveland and T.R. Jacobson (1992) "Overwintering by trilobite larvae of the horseshoe crab Bull. Cons. Biol., 157: 494-505

Castro, G., J.P. Meyers, and A.R. Place (1989) "Assimilation Efficiency of Sanderling (Calidris alba) feeding on horseshoe crab (Limulus polyphemus) eggs" Physiol. Zool., 62:716-731

Clark, K.E., L.J. Niles and J. Bager (1993) "Abundance and distribution of migrant shorebirds in Delaware Bay" Condor 95:694-705

Corven, J. (1998) "Shorebird Odysseys" Natural History 5:44-47

Fisheries Management Plan for Horseshoe Crabs (DRAFT): (1998) Atlantic States Marine Fisheries Report (Aug.); ASMF Commission pg. 17-20

Harrington, B. and C. Shuster (1999) "Crab Crisis at Delaware Bay" in: Defenders Summer pp:31-35

Horton, T. (1996) "Baiting the Blue Bloods" Audubon p:76-87

Jegla, T.C. and Costlow, J.D. (1982) "Temperature and salinity effects on developmental and early posthatch stages Limulus polyphemus on a sandy beach of Delaware Bay" Mar. Ecol Prog. Ser., (88): 289-292

Niles, L. (2000) "Red Knots" in: Field Notes, NJ Field Office, USFWS, pg. 10 (April) of Limulus" In: Bonaventura, J., Bonaventura, C. and Tesh, C. (Editors) "Physiology and biology of horseshoe crabs: studies on normal and environmentally stressed animals", Liss Publishers, NJ, pp. 103-113

Prior, R.B. (Editor) (1990) In: "Clinical Applications of the Limulus Amoebocyte Lysate Test" CRC Press "The American Horseshoe Crab: Limulus polyphemus" by C.N. Shuster Jr., Chap. 2: pp 16-25

Rudloe, A. (1979) "Locomotor and light responses of larvae of horseshoe crabs, Limulus polyphemus (L)" Biol.

Rudloe, A. (1980) "The Breeding behavior and patterns of movement of horseshoe crabs, Limulus polyphemus.L. in the vicinity of breeding beaches in Appalachia Bay, Florida" Estuaries:3;177-183.

Shuster, C. and Bottom, M. (1985) "A contribution to the Population Biology of horseshoe crabs: Limulus polyphemus (L) in Delaware Bay" Estuaries: 8(4): 363-372

Ward, P.D. (1991) "On Methuselah's Trail: Living Fossils and the Great Extinction" ISBN-0-7167-2203-8 pp. 135-150.

PART II

Biology and Evolution

Photo by Christine Kurtzke.

"Animals are far more fundamental to our thinking than we supposed.
They are not just a part of the fabric of thought: they are part of the loom"
in *Dreaming Elands*, Steinhart (1989: 816).

Chapter 2

Two Perspectives: Horseshoe Crabs During 420 Million Years, Worldwide, and the Past 150 Years in the Delaware Bay Area

CARL N. SHUSTER, JR.
Virginia Institute and School of Marine Science , The College of William and Mary

> *"Horseshoe crabs are ecological generalists:*
> *they can withstand a wide range of conditions, natural as well as man-induced". Their "...jack-of-all-trades nature... holds the clue to why they have remained so stable, so evolutionarily unchangeable for hundreds of millions of years."*
>
> Niles Eldredge, 1991, *FOSSILS. The Evolution and Extinction of Species*,
> Harry N. Abrams, Inc., Publishers, New York

The above quotation from one of Dr. Eldredge's books was selected to open this chapter because it succinctly ties what is understood about the biology and behavior of *Limulus polyphemus* to considerations of the evolution of horseshoe crabs. The quote also aptly serves to remind us to forego a too rigid application of what we find out about a population at a specific place and at a certain time to other horseshoe crabs throughout their geographic and geologic distributions.

Centuries ago, English scientists named this large marine invertebrate the "King Crab" (Fig. 1) and it was not until 1881 (Lankester) that horseshoe crabs were no longer considered to be Crustacea. That old appellation is a holdover from the colonial days of our country and is still used locally by watermen. Thus the early records describe the fishery for *Limulus* as the King Crab fertilizer industry.

Limulus in the Limelight, Edited by John T. Tanacredi
Kluwer Academic/Plenum Publishers, New York, 2001

Figure 1. Five "King Crabs" from the account by Fowler (1908, Plate 59: courtesy of the New Jersey State Museum, Trenton). The large, overturned female on the left is flanked by two males (the ventral view of one shows the claspers used in mating). The two crabs to the right are a mated pair (or placed to simulate amplexus).

The reader's attention is first directed to the changes in the body plan of horseshoe crabs over geologic time and then to the utility of that anatomy in the extant species, especially in swimming and burrowing. The fertilizer industry of Delaware Bay and its impact on the abundance of horseshoe crabs are then considered, followed by summary remarks on management of the crabs as a fishery resource. These topics could easily be expanded to book length treatises, so the following account is obviously brief.

1.   BODY FORM THROUGH GEOLOGIC TIME

One aspect of horseshoe crab evolution, their body form and its utility, can serve as a guide to their 420 million year 'journey." I will not dwell on how their unique body form might have served in the past and present survival of horseshoe crabs but characteristics of the present body form will be summarized.

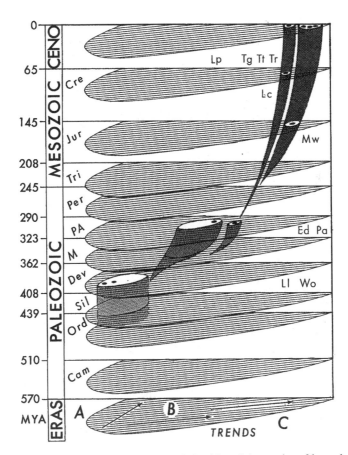

Figure 2. A conceptualized rendition of the relationships of the species of horseshoe crabs considered in this chapter, in terms of three general trends: A) the tendency for the opisthosoma to be shorter, B) articulating segments of the opisthosoma to become a fused unit, and C) the tendency to be larger (modified from Shuster, 1960). The intent is to suggest trends rather than to portray possible lineages. The geologic time scale is adapted from Eldredge (1991). Abbreviations: MYA = Million Years Ago. Geologic Periods: Cam = Cambrian, Ord = Ordovician, Sil = Silurian, Dev = Devonian, MS = Mississippian, PA = Pennsylvanian, Per = Permian, Tri = Triassic, Jur = Jurassic, Cre = Cretaceous. Fossil species: Ll= *Legrandella lombardii* (see Fig. 3); Wo = *Weinbergina optizi* (see Fig. 4); Ed = *Euproops danae* (see Fig. 5); Pa = *Paleolimulus avitus*; Mw = *Mesolimulus walchi*; Lc = *Limulus coffini*. Extant species (Fig 10): Lp = *Limulus polyphemus*, Tt = *Tachypleus tridentatus*, Tg = *T. gigas* and Tr = *T. (Carcinoscorpius) rotundicauda*.

The earliest species that have been recognized as belonging to the assemblage of animals called horseshoe crabs have been found in geologic formations of some 420 years ago. They had elongate bodies with up to twelve articulating segments and superficially looked as much like trilobites as the horseshoe crab types that followed. It is for this reason, in a brief account, that I will start the consideration of a "family tree" during the Devonian Period. (Fig. 2) Changes in body form, illustrated by a few selected species, have been used. For more complete analyses of the evolution of horseshoe crabs, the reader should start with Fisher (1984) and Anderson and Selden (1996).

2.   REPRESENTATIVE HORSESHOE CRABS

Even the body of early groups of horseshoe crabs was comprised of three recognizable parts: 1) a large cephalic shield (prosoma), 2) a mid-piece or trunk (opisthosoma) that had moveable segments in the earliest types but was fused into one piece in the more recent types, and 3) a single tail piece (telson).

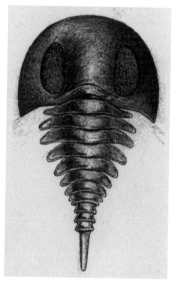

Figure 3. A diagrammatic representation of *Legrandella*, based on the description by Eldredge (1974). Prosomal width was about 6 centimeters. The opisthosoma consisted of 11 free, articulated segments; the first 8 segments were hemi circular with lateral extensions and the last 3 were cylindrical segments and supported a telson.

Two of the "older" species that lived during the Devonian (roughly 380 million years ago) had moveable opisthosomal segments. Fossils of *Legrandella lombardii* were found in Bolivia (Eldredge, 1974) (Fig. 3) and of *Weinbergina opitsi* in Germany (Stormer, 1955) (Fig. 4).

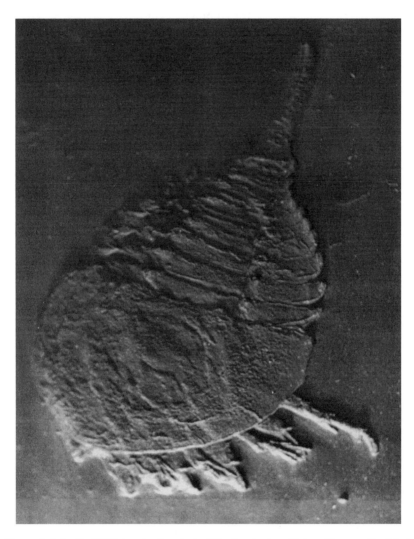

Figure 4. A fossil of *Weinbergina* (courtesy of the Senckenberg Nature Museum, Frankfurt-am-Main, Germany). This specimen was about 9 centimeters ling; the prosoma was relatively large; the "abdomen" consisted of 10 trilobite-like segments; the telson was short and stubby.

The more characteristic horseshoe crab body form was evident in *Euproops danae,* found in freshwater coal swamp habitats of Illinois in the Middle Pennsylvania (Fisher 1979) (Fig. 5). In *Euproops* the segments of the mid piece had fused yet were easily recognized in the fossils. Evidence of this segmentation still exists in the mid-piece of recent horseshoe crabs, as in the larva of *Tachypleus gigas* (Fig. 6). *Paleolimulus avitus* from the Middle Pennsylvania estuarine deposits of Illinois (about 310 mya) was among the first species to have the typical "modern" horseshoe crab body plan (Fig. 7).

Figure 5. Euproops danae (courtesy of National Museum of Natural History, Washington, DC). These were small animals that measured up to 45 millimeters wide at the opisthosoma. Relatively large numbers of this species, large in terms of the abundance of most horseshoe crab fossils, have been collected from ancient coal swamps that existed in Illinois during the Pennsylvania period. This specimen was undoubtedly much flattened during the fossilization process, particularly noticeable in the flaring of the terminal projections of the prosoma. Most of the axial ridge of the opisthosoma has broken. In other species it is a prominent feature.

The Upper Jurassic species, *Mesolimulus walchi* (Barthel, 1974), found in limestone deposits in Bavaria, Germany (about 150 mya), and *Limulus coffini* from Colorado (about 60 mya: Reside and Harris, 1952) were clearly horseshoe crabs in the "modern" form (Figs. 8 and 9, respectively).

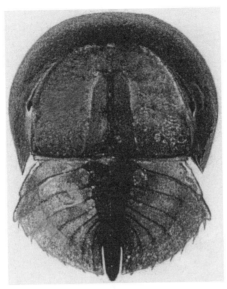

Figure 6. Larva of *Tachypleus gigas* (courtesy of Dr Kochi Sekiguchi). This larva is about twice the size of the smaller *Limulus polyphemus* larva that has a prosomal width slightly over 3 millimeters.

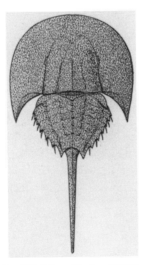

Figure 7. Paleolimulus avitus (modified from Dunbar (1923) and after examination of a few specimens). In many respects this fossil species appears like a tiny *Limulus polyphemus*. An average-sized specimen is about as big as *Limulus* in its first year of life in Delaware Bay (5th and 6th instars, with average prosomal widths of 12 and 17 millimeters, respectively: Shuster, 1979).

Figure 8. Mesolimulus walchii; body length of about 9 centimeters (courtesy Helmut Leich). This species was preserved in limestone deposits in Bavaria, Germany. This specimen preserves several features of the interior structures of the carapace, including the ventral rim of the prosoma, the ridges bearing the compound eyes, the axial ridge and the long moveable marginal spines of the opisthosoma. The elongated pit in the opisthosoma area is where the alimentary canal would have been in a live mammal.

There are four living species of horseshoe crabs, one along the Atlantic coast of North America and three in Indo-Pacific waters (Fig. 10). Each is commonly named according to an associated landmass. The behavior and ecology of the Asiatic species are of particular interest since they overlap in many areas in their distribution. All students of horseshoe crabs have benefited from the studies of Dr. Koichi Sekiguchi and his colleagues in Japan who not only have conducted extensive field studies throughout the Indo-Pacific but, among many studies, also have cultured and observed all four extant species, side-by-side, and have produced hybrids from certain crosses experimentally (see Sekiguchi, 1988).

Figure 9. The only record for the species *Limulus coffini* is this almost complete opisthosoma (courtesy of the National Museum of Natural History, Washington, DC). Its axial ridge was 6 centimeters in length. It is closer to the morphology of the opisthosoma of Limulus polyphemus than any other species, extinct or extant.

• *Limulus polyphemus,* the "American or Atlantic" horseshoe crab in along the coast of Yucatan and ranges from the Gulf coast of Florida to Maine. It is obviously, for Americans, the "reference species" introducing us to all others.
• *Tachypleus tridentatus,* the "Japanese" horseshoe crab, is the only species found in Japanese waters. There it is an endangered species, mainly due to the construction of dikes and polders in many coastal areas and pollution in others.
• *Tachypleus gigas,* the "Chinese" horseshoe crab is also found in the

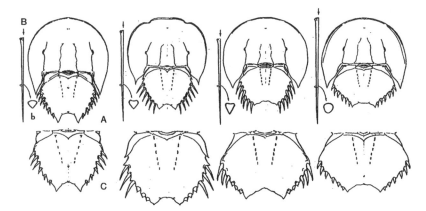

Figure 10. Diagrammatic comparison of the four extant species of horseshoe crabs, from left to right: *Limulus polyphemus, Tachypleus tridentatus, T. gigas,* T. (=*Carcinoscorpius*) *rotundicauda* (modified from Shuster, 1960). The top tier (sequence **A**) represents adult males, all drawn to the same prosomal width. Sequence **B** shows one half of each telson (cut along the axis) and sequence **b** shows the shape of cross sections of the telsons. Sequence **C** represents the sizes and shapes of the opisthosomas of the females, drawn to correspond to the males. While there is a considerable range in the sizes of adults in each species, *T. tridentatus* is usually as large or larger than Limulus. *T. rotundicauda* is the smallest of the four species.

Bay of Bengal and the Indonesian islands.

• *Tachypleus* (= *Carcinoscorpius) rotundicauda,* the *"Indonesian"* horseshoe crab that frequents mangrove swamps also ranges from the Bay of Bengal eastward into the islands.

On a short-term basis (at least over a period of a few decades), distinct populations of *Limulus polyphemus* can be identified on the basis of body dimensions (Shuster, 1955, 1979 and Riska, 1981). Many populations of *Limulus* are geographically discrete. However, they can travel sufficient distances that some populations have a wide distribution. In studies of distribution, the distances tagged animals have traveled appear to be largely related to the number of animals in a population (Shuster, 1950; Baptist *et al.,* 1957; Swan, 1996 pers. comm. (see Shuster, 1997). In the Delaware Bay area, the "home" of the greatest number of horseshoe crabs anywhere, a large part of the population that spawns on the shores of the bay comes in from the continental shelf. The impetus for their initial travel out of the bay as juveniles probably derives from their search for food.

On a geologic time scale it is evident from a study by Saunders *et al.* (1986) that there has been sufficient overlapping of populations for "gene flow" to occur. They demonstrated that there was a gradual change in mitochondrial DNA over the northern part of the latitudinal range of

Limulus, from Georgia northward. They found, however, a sharp demarcation between the populations on the Atlantic and Gulf coasts of Florida and those to the north. Saunders and her colleagues hypothesized that these differences accrued during a one millions year period. One interpretation of their study is that the Florida populations may have been cut off from those to the north by a physical barrier. Once the east and west coast Florida populations, now separated, were in contact via a large inland sea.

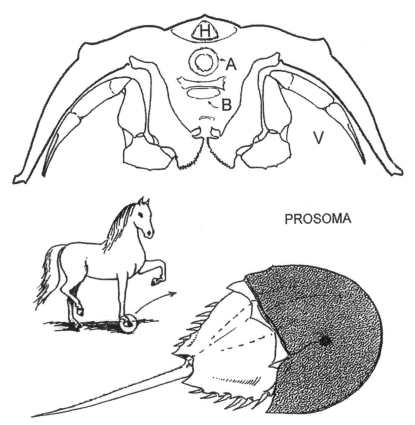

Figure 11. The highly arched, horseshoe-foot shaped front piece of the basic horseshoe crab is essential to many functions. Structures include: 1) a deep axial body which contains all of the vital organs [**A** = alimentary tract; **B** = brain; **H** = heart which lies within a sinus] and 2) a vault [**V**] or caisson within which the appendages can operate "protected" from outside disturbances.

3.   UTILITY OF THE HORSESHOE CRAB BODY FORM

The foregoing brief review of the geologic record of horseshoe crabs illustrates that the well-recognized body form has been around for a long time. At this point we are also ready to consider some aspects of the utility of this body form.

The highly arched, horse foot-shaped front piece of the body is essential to many functions, chiefly because it forms a "caisson" within which the appendages can work protected from outside disturbances (Fig. 11). The axial ridge of the extant species is elevated above the ridges bearing the lateral compound eyes. It forms the *"roof"* of the deep axial portion of the body that contains the largest volume of body parts (e.g., tubular heart, alimentary canal, nervous system) and the musculature of the bases of the appendages. These structures are embedded in connective tissue amid branches of the circulatory system, digestive gland, and the gonads. The latter tissues extend into the lateral portions of the body. The muscles of the bases of the legs are attached to the "roof" of the forepart (prosoma). This is seen externally by marking that break up the usual mosaic pattern of the outer shell between the furrows on each side of the axial ridge and the compound eyes.

Eight pairs of appendages are aligned along the deep axial portion of the prosoma (Fig. 12). The first seven (chelicerae, 5 pairs of ambulatory structures, and the chilaria) function in feeding. The eighth pair is fused and forms a broad protective covering (operculum) for the 5 pairs of branchial appendages (the "book gills") in the middle piece (the opisthosoma). Like the prosoma, the significance of the opisthosoma is in its deep-vaulted shape (Fig. 13). The vault has a strong rim that protects the branchia and is essential for the creation of water flow that is important in respiration, locomotion, and egg laying. Movement of the last six pairs of appendages creates water currents that flow under the crab, entering at the hinge area and exiting along side the telson. The jet-propelled "swimming" over the substrate, best described as a "scuttling" behavior, is another function enhanced by the deep-vaulted body.

The third component of the body of horseshoe crabs, the tip of the tail spine or telson can, has a wide excursion, up and down, and, limited by the terminal projections of the opisthosoma, moves about as far laterally as the width of the animal (Fig. 14). This movement is so that the tip of the telson can help over-turned crabs try to right themselves using the long telson. Mainly the larger, heavier animals use this righting process when they are stranded on their backs. The body is arched at the hinge with the telson pressed downward against the substrate (See Chapter 5, Fig. 8). The legs "kick" wildly causing an imbalance such that the crabs slips to one side or the other where the tips of the legs can gain a purchase against the sand and pull the crab over to its normal position. Smaller specimens and juveniles also use the telson to flip over but in water are often able to right themselves simply by jet propelling off the bottom and then flipping while swimming.

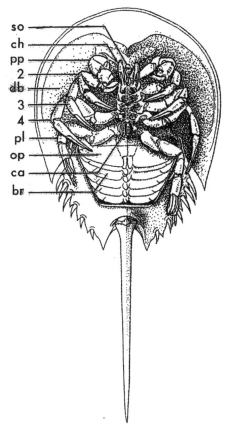

Figure 12. Ventral aspect of an adult male showing appendages and other structures: **so** = "sense organ"; **ch** = chelicera; **pp** = pedipalps or "claspers"), modified first pair of walking legs); **2** = second pair of walking legs; **db** = doublure (lining of the vault); **3** = third walking leg; **4** = fourth walking leg; **pl** = "pusher" leg; **op** = operculum; **ca** = chilarium; **br** = branchial appendages ("book gills").

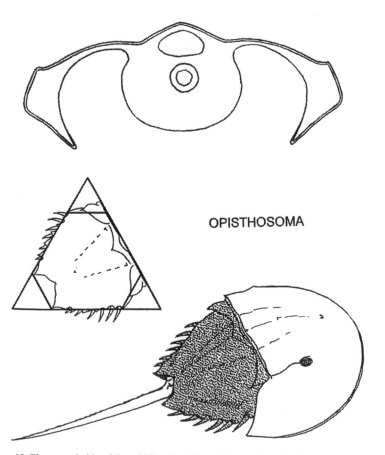

Figure 13. The ventral side of the middle-piece (the opisthosoma) of the horseshoe crab body has a deep vault. This vault has a strong rim and contains the five pairs of branchia ("book gills") that are essential for the creation of water flow that is important in respiration, locomotion, and egg-laying.

4.   A CONSIDERATION OF SWIMMING AND BURROWING

Limulus larvae are "programmed" to swim during embryonic development and, after hatching, can swim in calm water. They move in a jerky motion, moving upward when the appendages beat and sinking

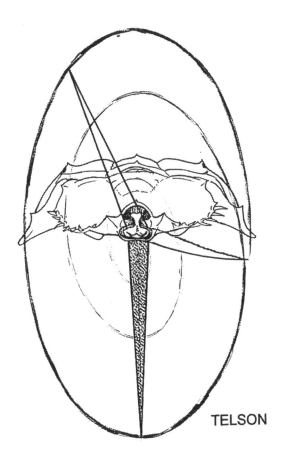

TELSON

Figure 14. The main direction of the telson is up and down. It is prevented from a wide lateral excursion by the terminal projects of the opisthosoma (see Fig. 13). The long telson is used by stranded, over-turned crabs to right themselves.

backward slightly in between strokes. Ascent in the water column is along a diagonal path until they reach the surface. There they are "held" in the meniscus unless bumped or if the water is disturbed, when they quickly sink back to the bottom. In aquaria, swimming is "promoted" by crowded conditions, especially with juvenile *Limulus*. As with the larvae, the final stage of swimming effort seems to be to reach the surface and plane out. At this time they look like swamped, miniature scows being propelled along by a somewhat metachronal beat of the appendages. Since calm waters are necessary for this stage of swimming, not many horseshoe crabs have been observed planning at the surface as they are easily "sunk". Beyond its possible use to get away from a beach after hatching and as a means of dispersal from crowded benthic or other conditions, there appears to be no other great utility or survival value in the ability to swim.

Burrowing is the dominant, more important, "every-day" function. Normally, burrowing (actually a process of digging in shallowly) is a prerequisite for, or at least an enhancement of feeding, resting, and egg laying. The deep vault of the carapace provides an important structural basis for this behavior.

5.   THE KING CRAB FERTILIZER INDUSTRY

The King Crab fertilizer industry of Delaware Bay is past history and only a few people remain who were involved in the fishery and fertilizer production. Yet it was a prominent commercial enterprise from at least 1850 to 1970 (Shuster, 1997 MS). Two points about this industry are pertinent to our discussion:

1.   How the crabs were taken and then processed into fertilizer, and;
2.   Changes in levels of abundance, based on early observations, commercial catches from 1870 to 1960, and observations, mostly on spawning activity, until 2000.

The earliest fishery for King Crabs probably was by the Native Americans, to use them and fish as fertilizer. According to Cook (1857), watermen and farmers were catching King Crabs in large numbers by the 19th century. At first the crabs were hand-collected from beaches during the spawning season. That practice is still in use today (as well as capture by dredge and trawl in some areas). It was the introduction of pounds along the New Jersey shore that greatly increased the harvests (Earll, 1887).

Smith (1891) described the construction of the pounds. These pounds were also used seasonally to catch fish. The leaders to the pounds extended out from the high-water line on the shore to the low-water level to a little beyond, of low tides. In some cases the pounds were several hundred yards or about 2,100 feet from the high-water line on the shore (Fig. 15). In essence the leaders "guided" the crabs when they were approaching the beach and then, when leaving after spawning, "led" them to the opening into the pound because they went bay ward as the water ebbed.

Over the years there have been several King Crab fertilizer plants in New Jersey and Delaware but not all of them were operating during the same period of years. In Delaware, the crabs were collected from the beaches by hand and taken to storage areas either in wagons or in scows. They were neatly stacked along the shores of tidal creeks (Fig. 16). In contrast, small quantities of the crabs were dumped behind the sand dunes along the New Jersey bay shoreline or, when large numbers were caught, they were tossed or dumped into pens (Fig. 17). After decaying and drying out, the crabs were taken to a fertilizer plant were they were pulverized by grinders, passed through heated passage ways to further dry the meal, then collected and bagged for distribution.

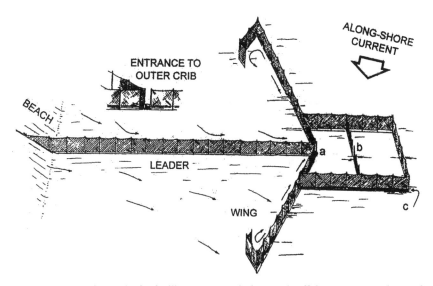

Figure 15. Horseshoe crabs for fertilizer were caught in pounds off the New Jersey shores of Delaware Bay from about 1870 to the 1960s. When the tide ebbed, horseshoe crabs that had been spawning left the beach and were "guided" by the leaders into the pen (crib) at the end of a pound (modified from Smith, 1891). Horseshoe crabs entered the pound via a ramp, falling off at the end into the crib (**a**). These pens were exposed at low tide. Fish were also caught in these pounds and, as the water ebbed, swam back into section **b** through narrow, vertical bars that prevented the crabs from entering the last crib. This separation of horseshoe crabs and fish kept the fish from being trampled at low tide. The pounds were fished daily at low tide, with the captured horseshoe crabs or fish being carried ashore via a skiff or horse-drawn wagon. The pounds were constructed of wire-covered poles. Extra wiring and horizontal slats (**c**) around the bottom of the pound net kept the crabs from digging underneath.

Figure 16. Thousands of neatly stacked horseshoe crabs on the shore of the St. Jones River, Delaware (from the file of agricultural photographs; courtesy the Delaware Archives). In the 1800s, a principle use of King Crab fertilizer in Delaware was in peach orchards.

Figure 17. Two views of a very large King Crab pen (from Fowler, 1908, Plates 64 and 65; courtesy of the New Jersey State Museum, Trenton). The mast of the boat serves as a reference point in both views. The numbers of horseshoe crabs has not been determined but, by a rough estimate, must total many more than ten thousand.

6. CHANGES IN ABUNDANCE

Shuster and Botton (1985) consolidated information, per decade, on horseshoe crab harvest from occasional annual harvest data in national fisheries catch statistics, from 1870's to the 1960's. Trends in abundance were also obtained from annual surveys of the spawning activity from 1990 through 2000 (annual summaries by Finn *et al.*, 1990 and Swan *et al.*, 1991-2000). Although the catch statistic and spawning survey data are not comparable, both data sets can be combined to illustrate major trends from 1870's to the present (Fig. 18). Even though the data are perceived as being non-rigorous (in a statistical sense) the graph does suggest that the Delaware Bay population of adult horseshoe crabs declined slowly during periods of heavy fishing pressure and, when the fishing lessened, slowly began to rebuild to a large populations. A delayed response to fishing, seen in the numbers of horseshoe crabs observed, is no doubt related to the size of the population and the extent of the fishing. Until better information is available we can postulate a general relationship between harvest level and abundance over time: 1) that the larger the population the slower the overall decrease and 2) that after a certain level of harvest has been reached, further harvesting reduces the population by 50% each decade. The lag time is also due, probably in a major way, to the lengthy life cycle of horseshoe crabs — the 10 to 12 years to reach maturity and the potential length of adult life up to 10 years. An increase in abundance began in the 1970's and continued to 923,000 adults reported in the 1990 spawning survey. Since the adult population was increasing, the expectation was that it would continue. It did not, falling to an average of 422,000 during the 1990's. Since there was no other obvious reason for the decline, the heavy harvesting for bait appears to have been the cause.

Successful, high levels of spawning activity, followed by high rates of survival of the juveniles yearly adds a strong year class to the spawning population. The probable sequence of events, after a spawning population is beat back year after year by increased fisheries activity, includes: 1) for the first few years (possibly up to ten) the recruitment of new spawners remains relatively high due to the former greater numbers of spawners, 2) ultimately the juvenile year classes are greatly reduced due to fewer spawners, 3) resulting in fewer young adults being added to the spawning population.

7. SUMMARY REMARKS - WHAT IS THE PROBLEM?

Of the animals that have been grouped together as horseshoe crabs, the

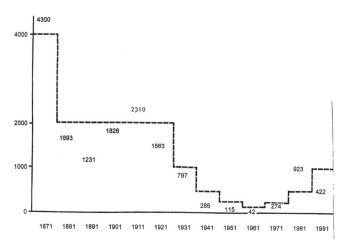

Figure 18. A rough estimate of the abundance of adult horseshoe crabs, *Limulus polyphemus*, in Delaware Bay from 1870 to 2000. The first year of each decade is given on the horizontal line and the number of horseshoe crabs (x 1,000) on the vertical line. Two sources of information have been used: 1871-1971 = commercial fisheries harvest data from federal fisheries agencies: 1979-2000, from spawning surveys. Due to the limited amount of data, the information for each decade has been averaged. The averages have been placed on the graph in the appropriate places. The dashed line emphasizes the general trend, decade by decade.

earliest members go back some 420 million years. By 310 million years ago, the unmistakable horseshoe crab body form was clearly established. The advantage of examining two entirely different kinds of information, the geological and historical, is that we have learned that horseshoe crabs are a persistent lot. Although none of the extant species had left a fossil record, their retention of the body form of "ancestral" types clearly justifies the claim that they are "living fossils." Historically, especially in the Delaware Bay area, the response of horseshoe crabs to heavy fishing pressures further confirms their ability to survive. We have not discussed survival in relation to environmental change, but in this too, horseshoe crabs are able to cope.

While there is no sound estimate of the peak numbers of *Limulus* that existed in Delaware Bay area at some time in the past, as in Colonial days, the adult fraction of that population must have included several millions, perhaps tens of millions. Otherwise it is difficult to conceive how over 4 million could be harvested in one year in the 1870's (Fig. 18). Throughout that time *Limulus* was undoubtedly, as we see today, plagued by other marine species ranging from microbes to mammals. We do not know to what extent their general "hardiness" and long life cycle conceal impacts by such "enemies" but spawning, certainly, has its visible risks -- many adults are stranded on beaches (Botton and Loveland, 1989). However, major impacts on the population have come from their harvest — for fertilizer in the past and for bait for eels, past and present, and the increasing whelk ("conch") fishery.

There is little doubt that a major change in the numbers of individuals in the horseshoe crab population could upset the ecological balance of the bay area. Over the past decade a reduction of nearly 50% may have occurred in the spawning population. A major question is whether a minimum number of spawning horseshoe crabs is essential for the success of the feeding/rest stop of the migratory shorebirds that stop at Delaware Bay on their way from South American wintering grounds to breeding grounds in Canada (see Clark *et al.*, 1993; Botton *et al.*, 1994; and Harrington and Shuster, 1999). If there is a strong linkage, which is most likely, then marked declines in the spawners is a matter of major concern. Declines in horseshoe crab populations, Delaware Bay and elsewhere, also impact upon the bait fisheries and the numbers of animals available for medical and research purposes. Biologists familiar with the situation believe the concern is real and that the horseshoe crab population of the Delaware Bay area should be better managed. This has led to a request from the Horseshoe Crab Management Board of the Atlantic States Marine Fisheries Commission for a public review of the situation. In response, the National Marine Fisheries Service published a proposed rule making (Federal Register Vol. 65, No. 200/Monday, October 16, 2000/ Proposed Rules, page 61135 — 50 Code of Federal Regulations Part 697) following the issuance of an environmental assessment in August 2000. The intent is to establish an area in Federal waters off the mouth of Delaware Bay as a sanctuary to protect the young adults of the spawning population. The most likely closure would represent an area equivalent to a radius of 30 nautical miles.

After more than a century of uncontrolled exploitation of a valuable, multi-use marine resource, a reasonable resource management plan appears to be evolving (Berkson and Shuster, 1999). And, despite all we know, much more research is needed to better understand horseshoe crabs and their role in nature and human commerce, for the benefit of all "users."

ACKNOWLEDGMENTS

In 1947, at Rutgers University, my interest in horseshoe crabs was sparked by Dr.Thurlow C. Nelson and in invertebrate paleontology by Dr. Helgi Johnson. Dr. Alfred Redfield at the Woods Hole Oceanographic Institution and Drs. Horace Stunkard and Harry Charipper of New York University, were important counselors as I began my horseshoe crab studies.

In more recent years, important biological associates have included Drs. William Hargis (Virginia Institute of Marine Science), Mark L, Botton, (Fordham University), Robert E. Loveland (Rutgers University), Robert B. Barlow Jr. (SUNY Syracuse, NY), H. Jane Brockmann (University of Florida) and Drs. Koichi Sekiguchi and Hiroaki Sugita (Tsukuba University). My pursuit of information on fossil species was aided by Drs. Niles Eldredge (American Museum of Natural History) Daniel Fisher (University of Michigan) and Lyle I. Anderson (University of Aberdeen. Scotland) and Ms. Jann Thompson (National Museum of Natural History, Smithsonian Institution, Washington, DC). In Germany my examination of *Mesolimulus walchi* and other fossil species has been assisted at museums and at private collections by: Dr. Günter Viol (Jura Museum, Eichstätt), Dr. Theo Kress (Maxberg Museum), Dr. Heinz Malz (Senckenberg Nature Museum, Frankfurt-am-Main), Helmut Leich in Bochum. B.A. Jürgen Schmitt in Frankfurt-am-Main, Georg Berger (Museum Berger) and Eduard Schopfel (JUMA, Gungolding-Altmühltal). Lastly, this article and Chapter 5 have benefited from my work with several of the above and other colleagues to prepare a manuscript for a scientific book to popularize *Limulus* and other horseshoe crabs.

REFERENCES

Anderson, L.I. and P.A. Selden. 1996. Opisthosomal fusion and phylogeny of Palaeozoic Xiphosura. Lethaia 30: 19-3 1.

Baptist, J., O.R. Smith and J.W. Ropes. 1957. Migrations of the horseshoe crab, *Limulus polyphemus,* in Plum Island Sound, Mass. U.S. Fish & Wildlife Service, Spec. Sci. Rept. Fish. No.220: l5pp.

Barthel, K.W. 1974. *Limulus:* a living fossil. Die Naturwissenschaften 61: 428-433.

Berkson, J. and C.N. Shuster Jr. 1999. The horseshoe crab: the battle for a true multiple-use resource. Fisheries 24(11): 6-10.

Botton, M.L. and R.E. Loveland. 1987. Orientation of the horseshoe crab, *Limulus polyphemus,* on a sandy beach. Biol. Bull. 173: 289-298.

Botton, M.L. and RE. Loveland. 1989. Reproductive risk: high mortality associated with spawning by horseshoe crabs *(Limulus polyphemus)* in Delaware Bay, USA. Marine Biol. 101: 143-15 1.

Botton, M.L., R.E. Loveland and T.R. Jacobsen. 1994. Site selection by migratory shorebirds in Delaware Bay, and its relationship to beach characteristics and abundance of horseshoe crabs *(Limuluspolyphemus)* eggs. The Auk 111(3): 605-616.

Clark, K.E., L.J. Niles and J. Burger. 1993. Abundance and distribution of migrant shorebirds in Delaware Bay. The Condor 95: 694-703.

Cook, G. 1857. Report on King Crabs in: Geology of the County of Cape May, State of New Jersey. Printed at they Office of the True American (Trenton): 105-112.

Dunbar, C.O. 1923. Kansas Permian insects. Part 2, *Pateolimulus,* a new genus of Paleozoic Xiphosura, with notes on other genera. Amer. J. Sci. V (30): 443-455.

Earll, R.E. 1 887. New Jersey and its fisheries. In: Goode, G.B. (ed.). The Fisheries and Fishing Industries of the United States. (U.S. Commission of Fish and Fisheries, Washington, DC), Section II: 397.

Eldredge, N. 1974. A revision of the suborder Synziphosurina (Chelicerata, Merostomata), with remarks on merostomes phylogeny. Amer. Mus. Novitates no. 2543:1-41.

Eldredge, N. 1991. Fossils. The Evolution and Extinction of Species. Harry N. Abrams, Inc. (New York): 100-Ill.

Finn, J.J., C.N. Shuster Jr. and B.J. Swan. 1991. *Limulus* spawning activity on Delaware Bay shores 1990. Finn-Tech Industries, Inc.: printed report (brochure).

Fisher, D.C. 1977. Functional significance of spines in the Pennsylvanian horseshoe crab *Euproops danae.* Paleobiology 3(2): 175-195.

Fisher, D.C. 1981. The role of functional analysis in phylogenetic inference: examples from the history of the Xiphosura. Amer. Zool. 21: 2 1-47.

Fisher, D.C. 1984. The Xiphosurida: archetypes of bradytely? In: Eldredge, N. and S.M. Stanley. Living Fossils. Springer-Verlag (New York): 196-213.

Fowler, H.W. 1908. The king crab fisheries in Delaware Bay. Annual Report, New Jersey State Museum, 1909 (Trenton): 111-119 + plates 59-65.

Harrington, B. and C.N. Shuster Jr. 1999. Crab crisis at Delaware Bay. Defenders 74(3): 3 1-35.

Lankester, E.R. 1881. *Limu/us* an arachnid. Quart. J. Microsc. Sci. 21:504-548, 609-649.

Reeside, J.B.Jr. and DV. Harris. 1952. A Cretaceous horseshoe crab from Colorado. J. Washington Acad. Sci. 42(6): 174-178.

Riska, B. 1981. Morphological variation in the horseshoe crab *Limulus polyphemus.* Evolution 35(4): 647-658.

Saunders, N.C., L.G. Kessler and J.C. Avise. 1986. Genetic variation and geographic differentiation in mitochondrial DNA of the horseshoe crab, *Limulus polyphemus.* Genetics 112:613-627.

Sekiguchi. K. (ed.). 1988. Biology of Horseshoe Crabs. Science House Co., Ltd. (Tokyo): 428pp.

Shuster, C.N.Jr. 1950. Observations on the natural history of the American horseshoe crab, *Limuluspolyphemus,* Woods Hole Oceanogr. Inst. Contrib. No. 564, 18-23.

Shuster, C.N.Jr. 1960. Xiphosura. McGraw-Hill Encyclopedia of Science and Technology 14:563-567.

Shuster, C.N.Jr. 1979. Distribution of the American horseshoe "crab," *Limuluspolyphemus* (L.). In: Cohen, E. (ed.) Biomedical Applications of the Horseshoe Crabs (Limulidae). Alan R. Liss, Inc. (New York): 3-26.

Shuster, C.N.Jr. 1982. A pictorial review of the natural history and ecology of the horseshoe crab *Limulus polyphemus,* with reference to other Limulidae. In: Bonaventura, J., C. Bonaventura and S. Tesh (eds.). Physiology and Biology of Horseshoe Crabs: Studies on Normal and Environmentally Stressed Animals. Alan R. Liss, Inc. (New York): 1-52.

Shuster, C.N.Jr. 1997. Appendix A: Annotated bibliography of horseshoe crab information. In: Farrell, J, and C. Martin. (eds.) Proc. Horseshoe Crab Forum: Status of the Resource. Univ. Delaware Sea Grant College Program (DEL-SG-05-97): 45-52.

Shuster, C.N.Jr. 1997 MS. The King Crab, *Limulus polyphemus,* Industry of Delaware Bay, from 1850 to 1970.

Smith, H.M. 1891. Notes on the king-crab fishery of Delaware Bay. Bull. U.S. Fish Commission 9(19): 363-3 70.

Stormer, L. 1955. Merostomata. In: Moore, R.C. (ed.) Treatise on Invertebrate Paleontology, Part P, Arthropoda 2: P4-P41.

Swan, B.J., W.R. Hall Jr. and C.N. Shuster Jr. 1991-2000. Annual reports on the survey of horseshoe crab spawning activity on the shores of Delaware Bay (processed).

Chapter 3

The Conservation of Horseshoe Crabs: What Can We Learn From the Japanese Experience?

MARK L. BOTTON

Department of Natural Sciences, Fordham University at Lincoln Center, NYC

1.   INTRODUCTION

There is a growing concern about the population status of the American horseshoe crab *Limulus polyphemus,* along the east coast of North America. This species is important commercially and ecologically, especially in the Delaware Bay area. Horseshoe crab eggs are an essential food source for migratory shorebirds during their northward (Spring) migration (Clark *et al.,* 1993, Botton *et al.,* 1994). Hundreds of thousands of shorebirds fly non-stop from South America to Delaware Bay during early May; they then feed intensively on horseshoe crab eggs to replenish their fat reserves in preparation for the next leg of their journey, which takes them to breeding grounds in the Canadian Arctic. Certainly, horseshoe crabs have proven to be extremely resilient over geological time, and their persistence since the Cretaceous is testimony to their ability to successfully cope with changing environmental conditions. Indeed, as Shuster (1982) has emphasized, *Limulus* is an ecological generalist, a species not tied to a specialized diet, bottom type, or temperature/salinity regime. It may be this adaptability to fluctuating environmental conditions that underlies its well-deserved reputation as a "living fossil." Why, then, are so many concerned about the survival of this species as we move into the 21st century?

The potential threats to *Limulus* populations are overfishing, the loss of prime spawning habitat because of erosion and/or coastal development, and pollution. Commercial fishing for horseshoe crabs is principally directed at adults (especially egg-bearing females) which are used as bait in fisheries for

Limulus in the Limelight, Edited by John T. Tanacredi
Kluwer Academic/Plenum Publishers, New York, 2001

41

eels and conchs *(Busycon carica* and *Busycotypus canaliculatum)*. By
conservative estimates, about 6.1 million pounds of horseshoe crabs were
landed in 1997, with the majority from New Jersey, Delaware, and Maryland
(Schrading *et al.*, 1998). Horseshoe crabs are also bled (non-destructively)
for the production of *Limulus* amoebocyte lysate (LAL), used by the
pharmaceutical industry for the detection of bacterial endotoxin (Novitsky,
1984). In the last century, a large commercial fishery in the Delaware Bay
area collected millions of horseshoe crabs for use as fertilizer and animal
feed (Botton and Ropes, 1987a). Estuarine sandy beaches, which are the
prime habitat for spawning by horseshoe crabs, have been adversely
impacted by coastal zone development, particularly the use of groins and
bulkheads in coastal properties. The political pressure to stabilize shorelines
is likely to increase as coastal erosion accelerates because of the sea level
rise associated with global warming. Finally, the propensity of horseshoe
crabs to spawn on estuarine beaches places their eggs in the proximity of
municipal and industrial discharges.

The habitat requirements and life-history characteristics of all four extant
species of horseshoe crabs appear to be rather similar (Shuster, 1982). The
basic premise of this paper is that information about the Japanese experience
with the decline and near extinction of *Tachypleus tridentatus* may provide
us with a number of valuable lessons as we strive to develop a conservation
strategy for *Limulus polyphemus* in North America. I will presume that such
a conservation strategy ought to do more than merely achieve the survival of
the species; to be successful, we need to insure that the population will
persist in numbers that ensure sufficient numbers of eggs to sustain migrant
shorebirds, and ideally, to permit a stable basis for the LAL industry and bait
fisheries.

2.   DISTRIBUTION OF THE INDO-PACIFIC HORSESHOE CRABS

Three species of horseshoe crabs reside in Asia. *Tachypleus gigas*, the
so-called Indian horseshoe crab, occurs widely from the Bay of Bengal
eastward into Southeast Asia, where it occurs in Thailand, Vietnam,
Indonesia, and elsewhere (Sekiguchi, 1988). Its range generally overlaps that
of *Tachypleus (Carcinoscorpius) rotundicauda*, and both species occur
together in places such as the Chonburi District of Thailand and on the
northern coast of Sumatra (Indonesia). Horseshoe crabs are considered to be
a delicacy in southeast Asia, but not in Japan. The "Japanese" horseshoe
crab, *Tachypleus tridentatus*, is distributed further to the north than the other
species. It ranges from the Indian Ocean coast of Sumatra north and east to

the Philippines, and along the Asian mainland from Indochina through China and Taiwan. The northern limit of its distribution is the western coast of Japan. In understanding the decline of the horseshoe crab in Japan and the possibility for recovery, it is important to note that its population is probably isolated from those in China, because of the vast distance of deep ocean separating them.

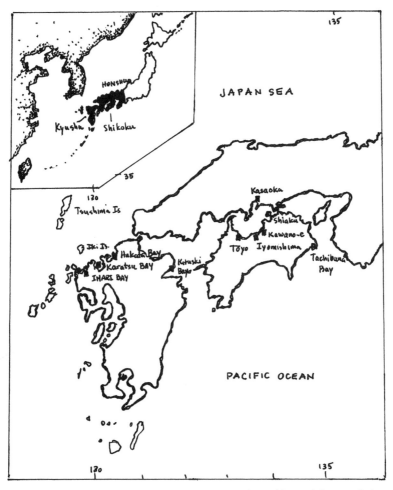

Figure 1. Map of Japan showing some of the major locations with remaining populations of horseshoe crabs. From Sekiguchi (1988)

Sekiguchi (1988) reviewed the historical and present distribution of *Tachypleus tridentatus* in Japanese waters. Its range, at least in historic times, appears to have been limited to the northern coast of Kyushu, and to the southern coast of western Honshu and the northern coast of Shikoku, which border the Inland Sea of Japan (Seto Inland Sea) (Fig. 1). There is consensus among Japanese scientists that the population throughout Japan has decreased precipitously during the last 50 or so years. Several local populations, particularly within the Seto Sea, have become extinct and others appear to be headed in that direction (Itow *et al.*, 1991, Itow, 1993). Since the Japanese horseshoe crab is neither eaten or used as bait, its decline cannot be attributed to commercial fishing. The two explanations most often cited by Japanese scientists for the decline of *Tachypleus tridentatus* are the loss of spawning beaches and pollution. During the summer of 1994, the author and Dr. Carl Shuster had the opportunity to visit a number of present and former breeding areas in coastal Kyushu and Honshu in collaboration with leading Japanese horseshoe crab biologists. We came away with the sense that although the "kabutogani" (literally, "helmet-crab") is revered as a symbol of Japan's efforts at marine conservation, habitat loss and pollution may have already done irreparable harm to the population.

3.   HABITAT LOSS

Japan is a densely populated country of some 125 million people, living in an island nation approximately 375,000 km^2 in area. Much of its mountainous interior is sparsely inhabited, and Japan's major population centers are located along the coast. To accommodate its growing population, many coastal areas have been enlarged by adding fill behind large concrete embankments or dikes adjacent to the water (Fig. 2). In some places, this has been done to extend agricultural production into areas that would have otherwise been intertidal beaches or marshes. Elsewhere, this practice of land reclamation has created industrial parks, shipping terminals, or land for commercial development or housing.

Nowhere has the impact of coastal development on horseshoe crabs been more dramatic or more devastating than in Kasaoka Bay, on the Honshu coast of the Seto Inland Sea. Horseshoe crabs once bred there in great abundance, and one particular area called Oe-hama Beach was set aside by the government in 1928 as a natural preserve and breeding area for horseshoe crabs (Sekiguchi, 1988). Notwithstanding this official designation, Oe-hama Beach along with much of the original Kasaoka Bay was reclaimed beginning in 1969. Ironically, most of the subsequent attempts at farming the reclaimed acreage failed because of seawater intrusion (H. Sugita, personal communication), and vast areas of the reclaimed land are now unutilized. In addition to destroying the breeding habitat at Oe-hama, the reclamation project also diverted the flow of the Yoshida River through a much smaller channel, thereby concentrating the waste discharges from Kasaoka City, upstream of the bay. At the present time, a sandy area along Kônoshima Channel is designated as a protected breeding area for the horseshoe crabs. However, it has been several years since any horseshoe crabs have been observed locally, possibly the result of poor water quality.

Figure 2. A portion of Imari Bay viewed from the sea wall at Tatara Beach. The extensive embankments have eliminated most of the natural spawning habitat for horseshoe crabs in this bay and elsewhere in coastal Japan.

Imari Bay, on the northern coast of Kyushu, was considered by Sekiguchi (1988) to have the largest population of horseshoe crabs in Japan. Even here, the abundance of horseshoe crabs is limited by the scarcity of suitable spawning habitat. One of the few areas for breeding is a protected area called Tatara Beach, which is a mere 16 m by 28 m patch located where a concrete wall and a breakwater meet at a right angle (Fig. 3). The site requires periodic replenishment to maintain the sand. A second sandy area of comparable size is approximately 0.1 km away. During 6 days of observations during the height of the1994 breeding season, Botton *et al.*, (1996) counted a total of 29 mated pairs at Tatara Beach, with a peak daily count of 5 pairs on August 7 and 8. In comparison, peak daily counts of between 20 and 30 mated pairs were recorded during surveys of Tatara Beach during the 1979 spawning season (Sekiguchi, 1988). Horseshoe crabs reportedly spawn elsewhere within Imari Bay, but much of the available shoreline has been diked. Of some encouragement was the fact that we were able to find viable eggs at Tatara Beach, and juvenile crabs on the tidal flats adjacent to the beach.

Figure 3. Tatara Beach is a protected area for horseshoe crabs located within Imari Bay in northern Kyushu, Japan. The entire beach is shown in the photograph. It is not uncommon for several hundred observers to visit the site every day during the breeding season in the hope of seeing the mating of kabutogani.

4.   EFFECTS OF POLLUTION

Although the loss of spawning habitat may be the major reason for the decline of the Japanese horseshoe crab, several studies have suggested that the susceptibility of eggs and embryos to pollutants is also a contributing factor in the decline or extinction of local populations. This is especially so along the heavily industrialized Seto Inland Sea, where Itow *et al.* (1991) reported that up to 20% of the embryos from some locations were malformed. The particular pollutant(s) responsible for the developmental abnormalities have not been identified, but in laboratory experiments with *Tachypleus* and *Limulus* eggs, Itow *et al.* (1998a) found that mercury, zinc, and organotins (TBT and TPT) were capable of inducing a variety of embryonic abnormalities when administered at levels ranging from 1 mg/L ($HgCl_2$ TBT) to 10-100 mg/L ($ZnSO_4$). Heavy metals, especially TBT, mercury, and cadmium, can also inhibit limb regeneration in *Limulus* larvae (Itow *et al.*, 1998b). The poor developmental success of the eggs in severely polluted regions of Japan poses an additional threat to the survival of the Japanese horseshoe crab.

5. PROSPECTS FOR THE RESTORATION OF JAPANESE HORSESHOE CRABS

Assuming that water quality was satisfactory and that sandy substrate was available, what would be the potential for a locally extinct population, such as Kasaoka Bay, to recover naturally? In marine species that recruit by means of widely dispersing planktonic larvae, such as oysters, hard clams, or mussels, the potential for natural repopulation via larval recruitment is much greater than it is in horseshoe crabs. Although the "trilobite" larvae emerge from beach sediments and enter the plankton, their dispersal capabilities appear to be quite limited; in fact, most horseshoe crab larvae appear to remain quite close to the breeding beach (M. Botton and R. Loveland, unpublished data). Recruitment could take place into Kasaoka Bay through adult migration; although there are no studies documenting long-ranging migration in *Tachypleus tridentatus,* I assume its capabilities are probably comparable to *Limulus* (Botton and Haskin, 1984, Botton and Ropes, 1987b). But, from all indications, the number of animals throughout the Seto Inland Sea is very small, so it seems unlikely that a significant number of mated pairs would migrate and spawn in Kasaoka Bay, unless the populations elsewhere in the sea increased as well.

Could aquaculture be a way to reestablish horseshoe crab populations? There is a small aquarium at the Horseshoe Crab Museum in Kasaoka City that has successfully reared horseshoe crab eggs obtained by artificial insemination, and the larvae and juveniles can be raised in a variety of systems (e.g. French, 1979, Kropach, 1979). However, *Tachypleus tridentatus* requires 15 to 16 molts (about 13 to 14 years) to reach sexual maturity (Sekiguchi, 1988), so it would be highly impractical to rear animals through adulthood. On the other hand, there is virtually no information about the rates of juvenile mortality in horseshoe crabs. It would be pure conjecture to predict the number of juveniles that would need to be released from a hatchery before we could anticipate a likelihood of successfully restoring a breeding population. A further complication is that *Tachypleus* is scarce throughout Japan, which would make it difficult to obtain the large number of eggs (at ca. 20,000 eggs/female (Shuster, 1982)) that would be needed for an aquaculture project to have a reasonable chance of success.

In Fall 1996, a team of scientists and local officials from Toyo City, locate on the island of Shikoku, visited the United States to discuss with Dr. Carl Shuster, Dr. Robert Loveland, Dr. Tomio Itow, and myself the prospects for replenishment of their horseshoe crab population. Toyo City has no horseshoe crabs remaining, in spite of the availability of sandy beach habitat and good water quality. One relatively low-cost suggestion was that *Tachypleus* eggs from other locations might be brought to Toyo City and reared within predator-protected beach sediments. But we cautioned that it might be many years before the success of such a project could be assessed.

6.   LESSONS FROM THE JAPANESE EXPERIENCE

The decimation of the Japanese horseshoe crab is clearly linked with the loss of spawning habitat, related to the practice of land reclamation and the construction of concrete enbankments and jetties along many of the former spawning beaches. Poor water quality, especially along the coast of the Seto Inland Sea, has adversely affected horseshoe crab egg survival on some of the remaining beaches. Consequently, *Tachypleus* has disappeared entirely from many locations where it was formerly abundant, and even the remaining populations are diminishing.

The basic reproductive biology of the Japanese and American horseshoe crabs is quite similar, and in particular, sandy estuarine beaches are preferred spawning habitats for both species (Botton *et al.*, 1996). Fortunately for *Limulus,* many important spawning beaches already receive protection. For instance, significant parcels of bay front property in Delaware Bay are under the control of the National Park Service or The Nature Conservancy. Breeding areas in Sandy Hook Bay, New Jersey and Jamaica Bay, New York are protected within the confines of the Gateway National Recreation Area. It is important to bear in mind, however, that global sea level and increased rates of beach erosion rise may drastically alter habitat quality in the next century. In the short term, erosion could expose peat sediments, which less suitable for egg-laying than sandy beaches (Botton *et al.*, 1988). Construction of bulkheads and similar structures makes an area less suitable for horseshoe crab spawning. At the present time, only about 10% of the Delaware Bay shoreline of New Jersey is judged to be optimal spawning habitat - about the same amount as has been disturbed by bulkheading (Botton *et al.*, 1988). The pumping of sand from offshore borrow sites onto beaches (beach nourishment) may be an ecologically preferable strategy to protect shoreline properties along the Delaware Bay shore (US Army Corps of Engineers, 1997).

In terms of pollutant effects, my sense is that *Limulus* populations have not been as severely impacted as the *Tachypleus* populations in Japan. Our laboratory has found that horseshoe crab embryos and larvae were highly tolerant of heavy metals (copper, zinc, and tributyltin) in comparison to similar stages in marine Crustacea (Botton, *et al.*, 1998a, b). Based on what is known about ambient concentrations of heavy metals in Delaware Bay and Sandy Hook Bay, it is unlikely that horseshoe crab eggs were suffering significant mortality due to these pollutants. Previous studies have shown that *Limulus* embryos and larvae were highly tolerant of pollution by oil (Laughlin and Neff, 1977, Strobel and Brenowitz, 1981), PCBs (Neff and Giam, 1977), and the pesticide diflubenzuron (Weis and Ma, 1987). Malformed *Limulus* embryos were very rare (<0.6%) in Delaware Bay and Sandy Hook Bay (Itow *et al.*, 1998a). The difference between the impacts of pollution on the American and Japanese horseshoe crabs may indicate a more serious pollution problem in Japan, a greater susceptibility of *Tachypleus* to pollution, or perhaps both.

Commercial fishing is the only pressure on the population that is more intense for *Limulus* than it is for *Tachypleus*. In principle, though, it should be far easier for us to legislate a remedy for overfishing than for the Japanese to undo the effects of land reclamation or coastal pollution. It has only been in the last few years that fishery managers have become aware of the magnitude of the horseshoe crab fishery and the trends that show a declining population. At the present time, state regulations on the fishery range from none (unrestricted) to various stipulations including permits, seasonal closures, gear limitations, and catch limits (Schrading *et al.*, 1998). We should take heed of the fact that the long lag between egg deposition and adult recruitment leads to the inevitable conclusion that once a horseshoe crab population is depleted, by whatever means, it may take many decades for recovery. Continued monitoring of the population and further scientific research may help to prevent the demise of this unique species.

ACKNOWLEDGMENTS

I would like to thank Dr. John Tanacredi for inviting me to participate in the mini-conference series. I am grateful to Drs. Hiroaki Sugita, Koichi Sekiguchi, and Carl Shuster for their friendship and insights during my trip to Japan in Summer, 1994. Drs. Tomio Itow and Robert Loveland provided invaluable help with the experiments on the sublethal effects of heavy metals. Lastly, my visit to Japan would not have been possible without the support of the Fordham University Research Council, the Japanese Society for the Preservation of Horseshoe Crabs, and the Horseshoe Crab Museum in Kasaoka City.

REFERENCES

Botton, M. L., and H. H. Haskin. 1984. Distribution and feeding of the horseshoe crab, *Limulus polyphemus*, on the continental shelf off New Jersey. Fish. Bull. 82:383-389.
Botton, M. L., M. Hodge, and T. I. Gonzalez. 1998a. High tolerance to tributyltin in embryos and larvae of the horseshoe crab, *Limulus polyphemus*. Estuaries 21:340-346.

Botton,   M. L., K. Johnson, and L. Helleby. 1998b. Effects of copper and zinc on embryos and larvae of the horseshoe crab, *Limulus polyphemus.* Arch. Environ. Contam. Toxicol. 3 5:25-32.

Botton, M. L., R. E. Loveland, and T. R. Jacobsen. 1988. Beach erosion and geochemical factors: influence on spawning success of horseshoe crabs *(Limulus polyphemus)* in Delaware Bay. Mar. Biol. 99:325-332.

Botton,   M. L., R. E. Loveland, and T. R. Jacobsen. 1994. Site selection by migratory shorebirds in Delaware Bay, and its relationship to beach characteristics and the abundance of horseshoe crab *(Limulus polyphemus)* eggs. Auk 111: 605-616.

Botton, M. L., and J. W. Ropes. l987a. The horseshoe crab, *Limulus polyphemus,* fishery and resource in the United States. Mar. Fish. Rev. 49:57-61.

Botton, M. L., and J. W. Ropes. 1987b. Populations of horseshoe crabs, *Limulus polyphemus,* on the northwestern Atlantic continental shelf. Fish. Bull. 85: 805-812.

Botton, M. L., C. N. Shuster, Jr., K. Sekiguchi, and H. Sugita. 1996. Amplexus and mating behavior in the Japanese horseshoe crab, *Tachypleus trideniatus.* Zool. Sci. (Tokyo) 13: 151-159.

Clark, K. E., L. J. Niles, and J. Burger. 1993. Abundance and distribution of migrant shorebirds in Delaware Bay. Condor 95:694-705.

French, K. A. 1979. Laboratory culture of embryonic and juvenile *Limulus,* p. 61-71. In: E. Cohen [ed], Biomedical applications of the horseshoe crab (Limulidae). Alan R. Liss, New York.

Itow, T., H. Sugita. and K. Sekiguchi. 1991. The decrease of horseshoe crabs in the Seto Inland Sea and the cause. Bull. Manage. Inform. Jobu Univ. 4:29-46. [In Japanese]

Itow, T. 1993. Crisis in the Seto Inland Sea: The decimation of the horseshoe crab. EMECS Newsletter 3:10-11.

Itow. T., T. Igarashi, R. F. Loveland, and M. L. Botton. 1998a. Developmental abnormalities in horseshoe crab embryos caused by exposure to heavy metals. Arch. Environ. Contam. Toxicol. 35:33-40.

Itow, T., T. Igarashi, M. L. Botton, and R. E. Loveland. 1998b. Heavy metals inhibit limb regeneration in horseshoe crab larvae. Arch. Environ. Contam. Toxicol. 35:457-463.

Kropach, C. 1979. Observations on the potential of *Limulus* aquaculture in Israel, p. 103-106. In: E. Cohen [ed.]. Biomedical applications of the horseshoe crab (Limulidae). Alan R. Liss, New York.

Laughlin. R. B. Jr. and J. M. Neff. 1977. Interactive effects of temperature, salinity shock, and chronic exposure to no. 2 fuel oil on survival, development rate and respiration of the horseshoe crab, *Limulus polyphemus,* p. 182-191. In: D. A. Wolff (ed.), Fate and Effects of Petroleum Hydrocarbons in Marine Organisms and Ecosystems. Pergammon Press, Oxford.

Neff, J. M. and C. S. Giam. 1977. Effects of Arochlor 1016 and Halowax 1099 on juvenile horseshoe crabs *(Limulus polyphemus),* p. 21-35. In: F. J. Vernberg, A. Calabrese, F. P. Thurberg and W. B. Vernberg (eds.), Physiological Responses of Marine Biota to Pollutants. Academic Press, New York.

Novitsky, T. J. 1984. Discovery to commercialization: the blood of the horseshoe crab. Oceanus 27:13-18.

Sekiguchi, K. 1988. Biology of horseshoe crabs. Science House Press, Tokyo

Schrading, E., T. O'Connell, S. Michels. and P. Perra. 1998. Interstate fishery management plan for horseshoe crab. Fishery Management Report No. 32, Atlantic States Marine Fisheries Commission, Washington DC.

Shuster, C. N. Jr. 1982. A pictorial review of the natural history and ecology of the horseshoe crab *Limulus polyphemus,* with reference to other Limulidae, p. 1-52. In: J. Bonaventura.

C. Bonaventura, and S. Tesh Feds.]. Physiology and biology of horseshoe crabs: studies on normal and environmentally stressed animals. Alan R. Liss, New York.

Strobel, C. J. and A. H. Brenowitz 1981. Effects of Bunker C oil on juvenile horseshoe crabs *(Limulus polyphemus).* Estuaries 4:157-159.

Titus, J. G. 1990. Greenhouse effect, sea level rise, and barrier islands: case study of Long Beach Island, New Jersey. Coastal Mgmt. 18:65-90.

Titus, J. G., R. A. Park, S. P. Leatherman, J. R. Weggel, M. S. Greene, P. W. Mausel, S. Brown, C. Gaunt, M. Trehan, and G. Yohe. 1991. Greenhouse effect and sea level rise: the cost of holding back the sea. Coastal Mgmt. 19:171-204.

U.S. Army Corps of Engineers. 1997. Villas & Vicinity, NJ Interim Feasibility Study. Draft Feasibility Report and Environmental Assessment. U.S. Army Corps of Engineers, Philadelphia District.

Weis, J. S. and A. Ma. 1987. Effects of the pesticide diflubenzuron on larval horseshoe crabs. *Limulus polyphemus.* Bull. Environ. Contam. Toxicol. 39:224-228.

Chapter 4

A Unique Medial Product (LAL) from the Horseshoe Crab and Monitoring the Delaware Bay Horseshoe Crab Population

BENJIE LYNN SWAN
Limuli Laboratories, Inc., New Jersey

1. INTRODUCTION: WHAT IS LAL?

The name Limulus Amebocyte Lysate was derived from: *Limulus polyphemus:* the Atlantic horseshoe crab; AMEBOCYTE: the circulating white blood cells; LYSATE: the breaking open or lysing of the white blood cells.

Limulus Amebocyte Lysate (LAL) is a medical product used to detect contamination in injectable drugs and medical devices. Clinically, LAL can diagnose diseases caused by gram-negative bacteria such as urinary tract infections, gonorrhea and endotoxemia. LAL can also be used to diagnosis gram-negative meningitis in just 15 minutes, whereas routine bacterial analysis of the spinal fluid may take 24 to 48 hours.

Bacterial contamination in injectable drugs and medical devices is referred to as pyrogens, derived from "pyro" meaning fire, or any substance that causes a fever. The majority of pyrogens in the pharmaceutical industry are endotoxins, high molecular weight proteins associated with the outer membrane of gram-negative bacteria. Endotoxins are everywhere in nature, especially abundant in water, and normally harmless to humans. The exception being when they come in contact with the blood stream and then they can be extremely toxic.

Limulus in the Limelight, Edited by John T. Tanacredi
Kluwer Academic/Plenum Publishers, New York, 2001

2.   THE HISTORY OF LAL

In the mid 1920's, studies conducted on rabbits proved that fevers associated with intravenous solutions were the results of bacterial products. This discovery enabled pharmaceutical companies to implement procedures to prevent contamination of their intravenous solutions. All drugs or devices that come in contact with the blood stream have to be tested for pyrogens. If the product contains pyrogens, a fever may result, followed by shock and possible death. All pharmaceutical products therefore have to be tested by the manufacturer before they are released into the market place. In 1942, the Food and Drug Administration (FDA) incorporated the first official rabbit pyrogen test in the United States Pharmacopoeia guidelines. The rabbit fever test was the standard test for endotoxins until the approval of the LAL test by the FDA in 1977. Companies tested their product by injecting the ear vein of a rabbit. The temperature of the rabbit was monitored for three hours. If the rabbit produced a fever, the product was contaminated. For many years, the rabbit test was the only practical pyrogen test, however the need for a simpler, more accurate test was recognized. (Pearson, 1985).

Figure 1. Testing rabbits for pyrogens

The discovery of *Limulus* Amebocyte Lysate started with a Bang! Dr. Frederik Bang to be exact, a pathologist at The Johns Hopkins University, is credited with the discovery of *Limulus* Amebocyte Lysate. Bang was studying the horseshoe crab's response to bacterial injections. His landmark study found that the horseshoe crab's blood had coagulated in the presence of the contamination and the product of the blood cells produced a gel in vitro when exposed to gram-negative bacteria. (Bang, 1953).

Dr. Jack Levin, a Fellow in Hematology at The Johns Hopkins University became interested in the crab's disease because it mimicked the course of human infection. After learning of Bang's observations, Levin started experiments concerning Limulus coagulation. Levin and Bang's research found that the amebocytes or white blood cells were responsible for the clotting and that endotoxin from the bacteria was the trigger (Levin and Bang, 1964). From the amebocytes, Levin developed a test reagent called Limulus Amebocyte Lysate (LAL) and began research on testing endotoxins in patients with serious diseases.

During this time, short-lived radiopharmaceuticals were emerging for use in diagnostic imaging and these new drugs had to be tested for pyrogens. However, radioactive drugs were not suitable for rabbit pyrogen testing because of their rapid decay, the large volume of the radioactive drug required for testing, radiation exposure and waste disposable problems. In 1969, James Cooper, a graduate student at The Johns Hopkins University under the direction of Levin and Dr. Henry N. Wagner, Chairman of the Division of Nuclear Medicine at The Johns Hopkins University began investigating the applicability of the LAL test to radioactive drugs and comparing the rabbit pyrogen test with the LAL test. Their study found a strong correlation between the rabbit's response and the LAL reactivity, indicating that the LAL test could be used for pyrogen testing (Cooper *et al.*, 1971). After this crucial discovery, Cooper, Dr. Ed Seligmann and Dr. Donald Hochstein with the United States Food and Drug Administration (FDA), conducted a study of 155 radiopharmaceuticals and biologicals, which established the feasibility of LAL as an alternative for endotoxin testing (Cooper *et al.*, 1972).

In 1977, draft guidelines were published for the use of LAL in testing blood products, vaccines, syringes and other disposable equipment. In 1987, the Food and Drug Administration issued official guidelines for pharmaceutical companies in order for them to use LAL in place of the rabbit test for release testing of drugs and medical devices. The lysate test is more commonly known in the pharmaceutical industry as the Bacterial Endotoxin Test (BET) and is the stated test in the United States Pharmacopoeia as well as the European Pharmacopoeia. In 1993, a revision to the Bacterial Endotoxin Test in the US Pharmacopoeia brought it in closer harmony with the LAL Test Guideline of the United States Food and Drug Administration.

The LAL test applications include the gel clot method, the chromogenic method as well as a turbidimetric test. The easiest method is the gel clot method. A small amount of LAL is added to an equal volume of the product to be tested. The mixed solution is then incubated for 1 hour at 37 degrees centigrade. If the product is contaminated, the mixture gels. If the product isn't, the mixture flows freely. The advantage of using of LAL over the rabbit test is less product needs to be used, less time is needed for the test and it is less costly. The development of *Limulus* Amebocyte Lysate was paramount in pyrogen or endotoxin testing in the pharmaceutical industry.

3.   LAL MANUFACTURE

Under the guidelines for LAL testing, all LAL manufacturers are required to be licensed by the U.S. Food and Drug Administration. In 1977, Associates of Cape Cod and Mallinkrodt became the first licensed LAL manufacturers. Presently, there are four companies in the United States licensed to manufacture this product. Charles River Endosafe that bleeds crabs in Charleston, South Carolina. BioWhittaker, Walkersville, Maryland has a bleeding facility in Chincoteague, Virginia and Associates of Cape Cod in Massachusetts and Haemachem, based in Saint Louis, Missouri with a bleeding laboratory in North Carolina.

The procedure for bleeding horseshoe crabs is relatively the same. The horseshoe crabs are collected live and brought into the laboratory. Only healthy, uninjured crabs are used, any crabs with the slightest injury are rejected. The crabs are cleaned and disinfected. Each crab is then immobilized by some type of bleeding stock and folded so the arthodial membrane (located in the center of the hinge) is exposed (Fig. 2). The membrane is swabbed with alcohol and a needle is injected. (The elongated heart of the horseshoe crab lies just under the axis of the shell). Blood is collected into an anticoagulant solution to protect the cells from prematurely opening until the cells are isolated. Once the blood is collected, the white blood cells are centrifuged or spun out to the bottom of the bottle. (A large crab yields as much as 12 ounces of blood, but only one ounce of white blood cells). The color of the blood is blue due to a copper binding protein, hemocyanin, in their blood that absorbs oxygen. The blue supernatent is decanted and the white blood cells are collected and then broken open or lysed. Inside the white blood cells is the product, termed lysate which can detect a certain level of bacteria but not sensitive enough for lower levels found in pharmaceuticals. The crude lysate is then further refined to make it

Figure 2. Bleeding a horseshoe crab

more sensitive , freeze dried for long term stability and packaged to be sold to pharmaceutical companies.

The horseshoe crabs are not killed by the bleeding process and it is a requirement of the United States Food and Drug Administration that the crabs be returned to their natural environment after a single collection of blood. Approximately, 200,000 to 250,000 horseshoe crabs are bled for the manufacture of lysate (Schrading *et al.*, 1998). A study by Ann Rudloe found bled horseshoe crabs have a 10% greater mortality than unbled crabs (Rudloe, 1983). However, mortality rates can vary for each bleeding facility and at Limuli Laboratories, Highs Beach, New Jersey, a lower rate (2.07%) was found (unpublished data).

4.   TAGGING AND SURVEY PROGRAMS

Since the bled horseshoe crabs are returned to the water, a tagging project was started in 1987 by James J. Finn to find out more about the horseshoe crabs longevity and migratory patterns. School groups as well as other groups became interested and approximately 30,000 horseshoe crabs have been tagged. The majority of the tagged horseshoe crabs have been released in New Jersey and Delaware with some tagging performed in Maryland, Virginia, New York and New Hampshire. Tagging usually occurred during the spawning season when the horseshoe crabs are most abundant along the beaches.

First, identifying information is recorded. The crab is sexed as to either male or female. This determination is performed visually by either comparing the size of the crab to other crabs (females are usually 25 to 30% larger than males), by examining the first pair of walking legs, which are modified in the male as "claspers" to aid the male in attaching to the female or by looking at the frontal arch of the carapace, which in the male is highly arched to fit over the female's opithosoma during spawning. The distance between the eyes are measured, termed the intraocular distance and sometimes the prosoma width is measured, the crab may be weighed and any abnormalities or encrusting organisms are noted. The age of the horseshoe crab may also be estimated by examining the appearance of the carapace.

Each tag is inserted into the carapace of the crab. A small hole is drilled into the trailing edge of the shell and a tag with a "Christmas Tree" pin is quickly inserted. The tagging can also be performed in the field by using a rechargeable drill. The drill bit may be outfitted with a rubber stopper to avoid drilling through both sides of the carapace.

The tag has a number to identify the crab as well as an address or telephone number for notification. A person reporting the tag will receive information on the project as well as the crab release information. A toll-free number (1-877-TAG-CRAB) was established in 1998 for reporting tag information. Some tags offer a reward to the finder of the tagged horseshoe crab.

Although the data accumulated during the tagging study has not been fully analyzed, it has confirmed some aspects of the crab's migratory behavior as well as longevity. Crabs that were tagged when spawning on the

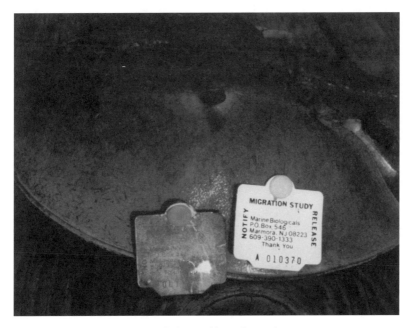

Figure 3. A tagged horseshoe crab

Delaware Bay shores have been recovered on the continental shelf, confirming that horseshoe crabs do enter and leave the bay. Horseshoe crabs tagged 1-2 miles offshore from Ocean City, Maryland in September 1998 moved onto Delaware Bay beaches the following spawning season. Crabs released from Delaware Bay beaches during the spawning season (May and June) were recaptured at locations along the continental shelf.

Horseshoe crabs released during the spawning season and found within days usually are found along the same side of the bay that they were released. After the horseshoe crabs have spawned and traveled away from the spawning beaches, there is a lesser chance they will be found in the same location of their release. After years of release, the horseshoe crabs are less likely to return to the same side of the bay that they were released. This evidence suggests that a relatively localized population exists in the vicinity of Delaware Bay although long distance migrations occur.

Horseshoe crabs are estimated to live at least 20 years, 10 years immature animals and the remaining years as adult spawners. Tagged crabs have been found years after they have been tagged. The most impressive return was from an animal tagged in 1988 and recaptured in 1996, 8 years later. This evidence suggests an age of at least 20 years for a horseshoe crab.

Shortly after the tagging project was started, a second project developed, an annual horseshoe crab spawning count or survey. The survey of horseshoe crabs was started in 1990 by the late James J. Finn and has continued with the assistance of Dr. Carl N. Shuster, Jr., an expert on the Atlantic horseshoe crab and Dr. William Hall, from the University of Delaware Sea Grant program and enormous help from numerous volunteers.

The basic idea of the horseshoe crab survey is that the adult population can be surveyed when the crabs come to the beach to spawn. Counting the crabs can be performed along the water's edge on the spawning beaches. At first, the concentration of effort was on determining if a volunteer force could adequately survey the horseshoe crab spawning population. Within the past few years, the concentration has been on obtaining the best possible data.

The survey has recently been modified to incorporate the needs of the Atlantic States Marine Fisheries Commission (ASMFC) Interstate Fishery Management Plan for the Horseshoe Crab, *Limulus polyphemus*. This plan was developed in response to localized declines in the horseshoe crab population and increasing fishing pressure due to the use of horseshoe crabs for eel and conch (whelk) for bait. One requirement of the plan is that a spawning survey be conducted yearly along the shores of Delaware Bay. The ASMFC Technical Committee opted to continue the Delaware Bay volunteer based survey with statistical guidance from the United States Geological Survey (USGS).

The objectives of the modified survey were to produce data more amenable to statistical analysis and to obviate weather related impacts and minor fluctuations in the behavior of crabs. One change included an increase in the number of days surveyed. The horseshoe crab count is conducted on twelve dates surrounding the four moon phases during the spawning season in May and June. The dates are comprised of two days before the new or full moon, the day of, and two days after the new or full moon. It was assumed that the five dates would profile spawning intensity at the most favorable tidal phase and thus 'capture' the peak in the number of spawners.

Survey beaches are randomly chosen from all the accessible beaches in New Jersey and Delaware. The process for choosing the beaches took into account the accessibility of the beaches, the historical spawning activity at each beach, and random selection. In 1999, 16 beaches were surveyed and in 2000, 20 beaches were surveyed during the season. Twelve beaches (6 on each side of the bay) will remain "fixed" beaches, meaning they will surveyed every year.

The survey is still a volunteer program but with the added assistance of the State Fisheries personnel. Most volunteers are "seasoned", they return year after year just like the crabs. Volunteers come from all walks of life, both federal and state employees help, different school groups, beach residents and some many others have contributed to the project. The volunteers are an important part of the project.

Some of the basic conclusions from the survey have confirmed earlier observations. More horseshoe crabs come at the night high tide. More males are found than females, which is behavioral aspect of their spawning. The males seem to hang around, whereas the females will spawn and leave the beach area. The survey has documented the spawning beaches along the bay shore and the associated levels of spawning activity since 1990. The majority of the spawners can be found in the mid range of Delaware Bay. Their abundance plotted against the axis of Delaware Bay creates a bell shaped curve.

The spawning estimates from the survey has demonstrated a sharp decline in the population since 1990 and 1991 when the estimate was roughly 1.2 million spawning individuals. With the added horseshoe crab regulations in the Mid-Atlantic States, it is hopeful that the population will remain at stable numbers.

5.   CONCLUSION

Horseshoe crabs have more to offer than anyone of us realize. Their contributions to science are phenomenal. The most important contribution of horseshoe crabs is the unique medical product, LAL, which has no replacement. Every human being at one time in their lives will benefit from horseshoe crabs. With increased awareness and further study, hopefully the horseshoe crab will survive for many, many more eons.

REFERENCES

Bang, Frederick B. (1953). "The toxic effect of a marine bacterium on Limulus and the formation of blood clots." Biol. Bull. 105: 447-448.

Cooper, IF., J. Levin and H.N. Wagner. (1971). "Quantitative comparison of in vitro *(Limulus)* and in vivo (rabbit) methods for detection of endotoxin." J. Lab. Clin. Med. 78: 138-148.

Cooper, J.F., H.D. Hochstein, and E.G. Seligmann, Jr. (1972). "The Limulus test for endotoxin (pyrogen) in radio pharmaceuticals and biologicals." Bull. Parenter. Drug Assoc. 26: 153 -162.

Levin,Jack and F. B. Bang. (1964). "The role of endotoxin in the extracellular coagulation of *Limulus* blood." Bull. Johns Hopkins Hosp., 115(3): 265-274.

Levin, Jack and F. B. Bang. (1964). A description of cellular coagulation in the *Limulus*. Bull. Johns. Hopkins Hosp., 115(4): 337-345.

Pearson, Frederick C. Ill. (1985). "Advances in Parenteral Sciences/2: Pyrogens; Endotoxins, LAL Testing, Depyrogenation." Marcel Dekker, Inc., New York and Basel.

Rudloe, Anne. (1983). "The effect of heavy bleeding on mortality of the horseshoe crab, *Limulus polyphemus,* in the natural environment." Jour, Invertebrate Pathology 42: 167-176.

Schrading, E., T. O'Connell. S. Michels and P. Perra. (1998). Interstate Fishery Management Plan for Horseshoe Crab, Fishery Management Report No. 32 of the Atlantic States Marine Fisheries Commission.

Photo by Christine Kurtzke.

"In the past it was possible to destroy a village, a town…even a country. Now it is the whole planet that has come under threat. This fact should compel everyone…from now on, (to face a basic moral consideration) it is only through conscious choice and then deliberate policy that humanity will survive."

Pope John Paul II, *The Ecological Crisis: A Common Responsibility.*

Neotropical migrant bird species feed on Horseshoe Crab eggs. Photo by Don Riepe.

Chapter 5

Tracks and Trails

CARL N. SHU~
Virginia Institut~

1.

Cent~
into the l~
first credit~
Limulus po~ ~~published~~ (Lockwood, 1870). In those bygone years, the Reverend Samuel Lockwood made his observations while walking the shores of Raritan Bay, New Jersey. Those observations might have pertained equally to Staten Island and Jamaica Bay of his time and, even today, his remarks are excellent background reading. Beyond this, the behavior of *Limulus* had not attracted much scientific attention until the last half of the 20th century.

Today the American horseshoe crab, *Limulus polyphemus,* is practically a household topic, at least along the eastern seaboard and especially in the mid-Atlantic States. This attention, in the media and in the political-social arena, has been primarily focused on the fate of large flocks of migratory shorebirds that stopover on the shores of Delaware Bay. These birds fly non-stop from South America on way to breeding grounds in the Arctic. They arrive famished at Delaware Bay and feed voraciously on the eggs of *Limulus* (Clark *et al.*, 1993; Botton *et al.,* 1994). The synchrony of the arrival of the birds and horseshoe crabs on the shores of Delaware Bay is a marvel of nature. How dependent the birds are upon horseshoe crabs and how many crabs are necessary to produce the quantities of eggs that sustain the birds are matters of much contention.

Limulus in the Limelight, Edited by John T. Tanacredi
Kluwer Academic/Plenum Publishers, New York, 2001

There is much to be learned even when horseshoe crabs are not in sight. Their tracks and trails in sand and mud give transitory evidence of their passing by. Even more amazing, look-a-like fossilized tracks made by earlier species have been "frozen in rocks" formed millions of years ago. The following discussion is intended as a guide to observing horseshoe crabs. It is in part a discussion of what to look for, a sort of "detective" approach, and on how you might be able to gauge the "health" of a spawning population and "sleuth" the meaning of their tracks and trails.

2.   AGE AND APPEARANCE

No one has been able, by a direct method, to determine the age of an adult horseshoe crab. We do not even know precisely how long it takes it to become an adult. It is generally accepted that it takes ten years to mature (Shuster, 1982). That may vary considering the environmental conditions associated with the wide latitudinal range of the species, from Yucatan to Maine. We know that the adults can live for at least eight years based upon tagging (Swan, 1997, pers. comm.; Swan *et al.*, 1998) and from the age of an epibiont, the slipper snail *Crepidula fornicata* (Botton and Ropes, 1988). So, it is probably not incorrect to estimate a total life span as some 20 years. In speaking about the age of an adult it is customary to say that it is in its first or eighth year, or it is a 1-year or 8-year old (adult).

Table 1. A subjective guide to the age of adult horseshoe crabs, *Limulus polyphemus,* based upon the appearance of their carapace and activity.

Appearance	Age category		
Adult Age (in years)	Young 1 to 3	Middle-Age 4 to 7	Old 7 to 10
Erosion of Carapace	Little or no erosion; clean and lustrous; Mating scars on females usually evident within a year	Moderate abrasion large areas blackened	Nearly or completely black; often almost worn through in spots
Epibionts	Few if any	Common	Usually present and old
Behavior	Most Active	Moderately Active	Generally Sluggish

Despite the lack of a sure method of aging, it is possible to look at a group of horseshoe crabs and estimate their relative ages, at least whether they are young or old. Signs to look for in "aging" are the result of horseshoe crab activity in an abrasive environment and the fact that the adults rarely molt if ever. Thus, whatever erosion takes place or whatever attaches to its shell are clues by which to age a specimen (Table 1). Examining the attached

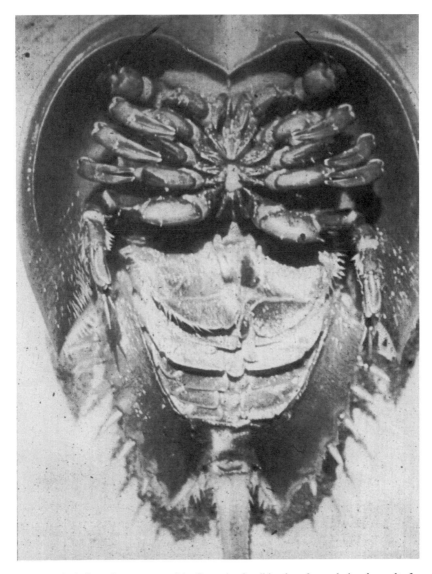

Figure 1. The bulbous last segment of the first pair of walking legs forms during the molt of a juvenile male into adulthood (unless otherwise stated, all photographs are by the author). This segment bears an enlarged, curved moveable claw (chela) and an atrophied fixed chela (designated by an arrow). It is the fixed chela that snaps off when a "virgin" male attempts to mate.

organisms is like looking at a miniature zoo. Because its carapace furnishes a substrate for so many epibionts, Allee (1922) described *Limulus* as a "walking museum" (see also Botton, 1981; Shuster, 1982).

The inherent value in examining the appearance of the adults may be that it might provide indirect evidence about the "health" of the population. If a spawning population is comprised of a large percentage of young animals, we can assume that it is either due to an unusual mortality among the older members of the population, a really "healthy" proportion of young active spawners, or a combination of factors.

Although an anatomically "virgin" adult male is a rarity, it is worth describing (Fig. 1). When a juvenile male molts for the last time into its mature form several changes have occurred: the forward edge of the prosoma is more highly arched and its first pair of walking appendages have been modified into claspers. These claspers are use to hold onto a female (amplexus). The "virgin" condition of these claspers exists only as long as the non-moveable claw, which atrophies on the last molt, has not yet been broken off. Just as soon as the male grasps a female this chela is snapped off, if not already lost due to other pressures as in feeding.

Appreciation of the behavior of *Limulus* during spawning can be gained from relatively recent research. (Botton *et al.*, 1988; Brockman, 1990, 1992, 1996; Brockmann and Penn, 1992; Penn and Brockmann, 1994, 1995; Shuster and Botton, 1985). A special portion of the spawning population, those stranded after spawning, furnishes another insight into the "health" of the population (Botton and Loveland, 1989). Many of the animals stranded on beaches after spawning are individuals, regardless of age, that have short telsons; others are the older and more feeble individuals.

Contrasts in the "age" appearance of adult horseshoe crabs are illustrated in Figures 2-4. It is easy to pick out the young, lustrous, and cleaner-shelled individuals in any assemblage, whether they are among spawning animals or among captives by trawl vessels.

3.   TRACKS AND TRAILS

The terms "track" and "trail" have been in use for one hundred years (Packard, 1900). Packard contributed several papers on the morphology and development of *Limulus* as well as on fossil Xiphosurans. He defined a "track" as an individual footprint and a "trail" as a series of footprints made by an individual. These definitions were accepted by Caster (1938) who made what is perhaps the most comprehensive comparison of extant and extinct Xiphosuran tracks and trails. Basically, there are three major sources of marks in a substrate: 1) those made by the appendages, 2) by the edges of

Figure 2. A close-up of a spawning scene at Jamaica Bay, New York, in May 1997. Note the differences in the appearance of the carapaces, seen in this illustration mainly in shades of gray and black -- light to darker shells. Arrows point to the young specimens (see Table 1).

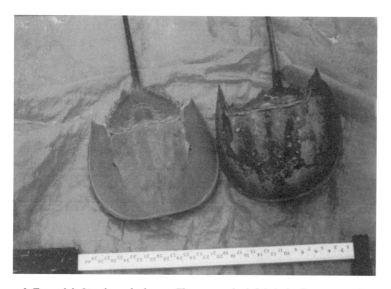

Figure 3. Two adult *Limulus polyphemus*. The one on the left is in its first or second year as an adult (i.e., 10 or 11 years old). The other is approaching old age (17 or 18 years old) -- its carapace is so severely eroded that only a thin shell exists in large sections.

Carl N. Shuster Jr.

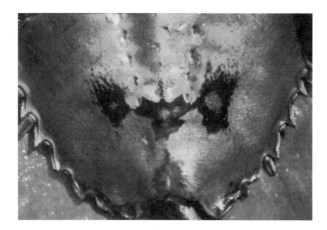

Figure 4. A dorsal view of the mid-section (opisthosoma) of a female in probably her second adult year (12+ years of age). There are several mating scars: 1) three prominent blackened area indicate at least two separate matings -- where the forward edge of the male carapaces abraded her shell; 2) a smaller abraded area is in the terminal axial position; it probably resulted from the rear-most part of the ventral shelf of a mate; and, 3) imprints of the male claspers have worn away part of the terminal projections of her opisthosoma.

the carapace, and 3) by the telson. Each of these may be in evidence to varying degrees in any one trail. What and how much that is recorded depends not only upon the substrate but also upon whether a trail was made when "crawling" on a beach or when "walking" or "gliding" across a submerged "muddy" flat. Less prominent marks may be made by the ventral, lateral rims of the front part of the body (the prosoma) and the moveable spines on the mid-body (opisthosoma).

Prints may be made by the appendages. There are two major footprints; those made by the four pairs of pincer-tipped walking legs and those by the heavier last pair of legs, the "pushers." Of these imprints, those of the "pusher" are more prominent and prevalent. Part of the crawling technique utilized by *Limulus* is to lift itself off the substrate and then lunge ahead. Lifting done with the walking legs while the pushers kick backwards. These dual actions create a slight "bobbing" and "lurching" motion as a crab moves down a beach.

After spawning on a beach, each horseshoe crab seems to have its own solution on the way to get back to the water. Essentially, *Limulus* uses the slope of a beach as a guide. It becomes disoriented on a flat area where its circuitous trails (Fig. 5) betray its lack of orientation (Botton and Loveland, 1987).

Figure 5. The circuitous trail of a disoriented horseshoe crab on an intertidal flat at Toms Cove, Chincoteaque, Virginia. Here, the slope of a beach (Botton and Loveland, 1987) the usual "guide leading a horseshoe crab toward water, is lacking. In this situation the circuitous routes may be a "search" pattern. If the animal does not give up and dig in, such a "search" sometimes does lead an animal back to water.

Prints are also made by the edges of the carapace. When *Limulus* is in contact with the substrate, as when crawling on a sandy beach, the forward portion of the carapace, acting like a plow, throws sand to the sides. This sand tends to pass along the sides of the carapace forming furrows. These, in conjunction with the furrow usually made by the telson, leaves a 3-rail trail (Fig. 6). The prominence of these furrows depends upon the moisture content and sediment particle size.

When digging in after being stranded by an ebbing tide on an intertidal flat (Fig. 7), a horseshoe crab usually plows up such a quantity of substrate that its carapace is almost covered (Rudloe and Rudloe, 1981). Such excavations made by *Limulus,* either to "rest" during low tide or to feed (Shuster, 1982), differ from those made by skates and sea gulls. Horseshoe crab excavations usually have one side broken down; skates have an "entrance" and exit" to their excavations; gulls leave circular puddles that are more completely ringed.

Prints may be made by the telson. The telson is a versatile organ that can perform several functions. Its actions are enhanced by its long length and almost universal joint connection with the mid-part (opisthosoma). The length is important in the righting process. Overturned crabs, whether stranded upside down on a beach or rolled over in the water by a wave, use

Figure 6. A female horseshoe crab heads for the water from the top of a beach. This specimen was one of several left stranded after spawning. After the water had ebbed sufficiently, it and the others were separately righted and their activity/behavior down the beach slope observed and photographed. This trail is one of several variations often observable. Typically the trail, as here, is "railed". It is formed by the edges of the prosoma creating furrows and, in this case, a broad "middle rail" created by the telson moving slightly from side-to-side. The spaces between the two outer and middle rails are punctuated by the impressions made by the large hind, "pusher" legs.

Figure 7. Flexing the body so that it arches up off the sand, is the first step usually taken by a stranded, over-turned horseshoe crab. Then while kicking vigorously with its legs the crab is unbalanced and tilts to one side until the edge of the prosoma contacts the sand. In that position, one or more of the thrusting legs is able to dig into the sand, enabling the animal to right itself.

the telson to right themselves. If the telson is too short, less than one-half the normal length, the crab has difficulty in arching sufficiently off the substrate to effectively roll over. The overall motion is that of forming an arch by flexing of the body at the hinge and pressing downward with the telson. This elevates the crab off the substrate. Then, by threshing and kicking with its legs, it becomes unbalanced and rolls to one side. With an edge of the carapace on the sand, the kicking of the legs usually brings them into contact with the sand. In time the legs are able to "grab" sufficiently into the sand to complete the turnover process. So, one of the telltale prints of the telson results from its thrusting about when the animal is overturned (see Fig. 8).

Truly fascinating are the resultant fossilized trails of horseshoe crab species. The most famous trails were made by the species *Mesolimulus walobi* that lived some 145 million years ago. Its tracks were recorded in the Jurassic limestone formations in what is now Bavaria, Germany. Of the several collections of fossils that I have seen, the one at the Museum beim Solenhofer Aktien-Verein, Maxberg, Germany (courtesy of Dr. Fheo Kress, Director) is perhaps the single best to visit. There are a number of tracks and trails, but the most photographed trail (Fig. 9) shows a disoriented animal (see Malz, 1976 pages 54-56, Rudloe and Rudloe, 1981 page 566, and the discussion by Barthel, 1974).

Figure 8. Sometimes neither flexing the body nor pressing their telson against the beach are sufficient to elevate the crab up to a "roll-over" position. Yet the crab may still thrust with its telson and tilt slightly so that it comes to rest just a few inches from its former position.

Figure 9. In this scene a stranded, overturned adult horseshoe crab has unsuccessfully tried to right itself many times. The tell-tail marks surrounding crab were formed by the pressure of the posterior part of the telson against the sand. What makes this an unusual observation if the 360-degree rotation that the crab made during the attempts to right it. The number of attempts can be counted by the depressions in the sand (photograph by M. Benjie Lynn Swan)

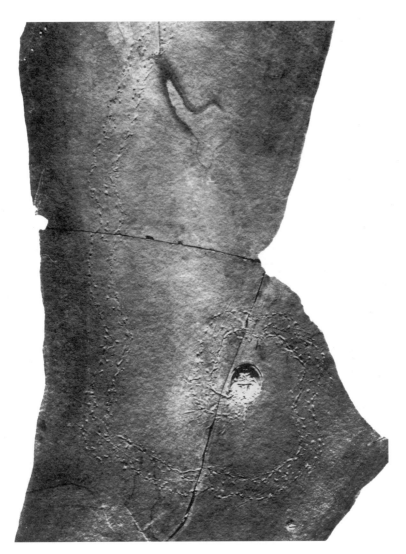

Figure 10. A fossilized trail of a horseshoe crab is preserved on this large slab of lithographic limestone taken from a quarry in Bavaria, Germany (courtesy of Dr. Theo Kress). The trail begins at the top, center of the slab and moves down near the left side of the slab. Near the bottom, the animal veered to the right on an ever-decreasing circular path with the animal preserved at the end. This is an example of the pathway of a disoriented, stressed Jurassic horseshoe crab species *Mesolimulus walchi*.

A great number of tracks and trails have been found in a shale formation in Pennsylvania (Fig. 10). These trails were made by an earlier Xiphosuran, possibly *Protolimulus,* during the Upper Devonian. These were beautifully illustrated and masterfully interpreted by Caster (1938). In his intensive study he compared plaster of Paris casts of *Limulus* trails with the fossil trails. It is an excellent example of how the functioning of anatomical parts can be correlated with the marks made in the formation of the trails. His observations should be consulted by anyone interested in *Limulus* trails.

4. CONCLUSION

In summary, in sleuthing the trails of horseshoe crabs, whether observing spawning or in interpreting the trails they have left behind, a camera is a prerequisite for a complete record. No spawning group or any trail appear to be exactly alike another although there may be several similarities. That is what makes the deciphering so interesting; even when watching spawning or a trail being made. Afterwards, with only the trail remaining, much has to be imagined. Since the trails are so transitory, a good photograph is needed to refresh ones memory.

At least two main considerations must be made when looking at a trail. The characteristics and condition of the substrate and the various evidence of activity left by the crab. It is the attempt to unravel of all the possible combinations of conditions and activity that make the study of these prints so fascinating.

ACKNOWLEDGMENTS

The author thanks Dr. Theo Kress (Museum beim Solnhofen Aktien, Maxberg, Bavaria, Germany) for courtesies extended – especially the opportunity to examine and photograph many specimens of the Jurassic horseshoe crab, *Mesolimulus walchi* (see Fig. 10). He also wishes to thank M. Benjie Lynn Swan (Limuli Laboratories, Cape May, New Jersey) for the use of her photograph of the righting activity of an overturned adult (Fig. 9). All other photographs were taken by the author.

REFERENCES

Allee, Warder C. (1922). Studies in marine ecology II: An annotated catalog of the distribution of common Invertebrates of the Woods Hole littoral. Manuscript in the Marine Biological Library, Woods Hole, Massachusetts.

Barthel, K. Werner. (1974). *Limulus:* a living fossil. Die Naturwissenschaften 61: 428-433.

Botton, Mark L. (1981). The gill books of the horseshoe crab *Limulus polyphemus)* as a substrate for the blue mussel *(Mytilus edulis)*. Bulletin New Jersey Academy of Science 26: 26-28.

Botton, Mark L. and Robert E. Loveland. (1987). Orientation of the horseshoe crab, *Limulus polyphemus,* on a sandy beach. Biological Bulletin 173: 289-298. 1989. Reproductive risk: high mortally associated with spawning by horseshoe crabs *(Limulus poyphernus)* in Delaware Bay, USA. Marine Biology 101: 143-151.

Botton, Mark L., Robert E. Loveland, and Timothy R. Jacobsen. (1988). Beach erosion and geochemical factors: influence on spawning success of horseshoe crabs *(Limulus polyphemus)* in Delaware Bay. Marine Biology 99: 325-432. 1994. Site selection by migratory shorebirds in Delaware Bay, and its relationship to beach characteristics and abundance of horseshoe crab *(Limulus polyphemus)* eggs. The Auk 111: 605-616.

Botton, Mark L. and John W. Ropes. (1988). An indirect method for estimating longevity of the horseshoe crab *(Limulus polyphemus)* based in epifaunal slipper snails *(Crepidula fornicata)*. Journal Shellfish Research 7: 407-412.

Brockmann, H. Jane. (1990). Mating behavior of horseshoe crabs, *Limulus polyphemus.* Behavior 114: 1992. The effects of age on male behavior in *Limulus polyphemus.* Galaxea 11: 61-62.

Brockman, H. Jane. 1996. Satellite male groups in horseshoe crabs, *Limulus polyphemus,* Ecology 102: 1-21.

Brockmann, H. Jane and Dustin Penn. (1992). Male mating tactics in the horseshoe crab, *Limulus polyphemus*. Animal Behavior 44: 653-665.

Caster, Kenneth E. (1938). A restudy of the tracks of *Paramphibius*. Journal Paleontology 12: 1-60.

Clark, Kathleen E., Lawrence J. Niles and Joan Burger. (1993). Abundance and distribution of migrant shorebirds in Delaware Bay. Condor 95: 694-705.

Eldredge, Niles. (1970). Observations on the burrowing behavior in *Limulus polyphemus* (Chelierata, Merostomata), with implications on the functional morphology of trilobites. American Museum Novitates 2436: 17pp.

Lockwood, Samuel. (1870). The horseshoe foot crab. American Naturalist 4: 257-274.

Malz, Heinz. (1976). Solnhofener Plattenkalk: Eine Welt in Stein. Museum beim Solenhofer Aktien Verein, Maxberg, Germany: 109 pages.

Packard, Alpheus Springer. (1900). On supposed merostomatous and other Paleozoic trails, with notes on those of *limulus*. American Academy of Arts, Proceedings 36: 63-71.

Penn, Dustin and H. Jane Brockmann. (1994). Nest-site selection in the horseshoe crab, *Limulus polyphemus*. Biological Bulletin 187: 373-384. (1995). Age-biased stranding and righting in horseshoe crabs, *Limulus polyphemus.* Animal Behavior 49: 153 1-1539.

Rudloe, Anne and Jack Rudloe (photographs by Robert Sisson). (1981). The changeless horseshoe crab. National Geographic 159: 562-572.

Shuster, Carl N. Jr. (1982). A pictorial review of the natural history and ecology of the horseshoe crab *Limulus polyphemus,* with reference to oilier Limulidae. In: Bonaventura, Joseph, Celia Bonaventura, and Shirley Tesh (eds.), Physiology and Biology of Horseshoe Crabs: Studies on Normal and Environmentally Stressed Animals. Alan R. Liss, Inc.: 1-52.

Shuster, Carl N. Jr. and Mark L. Botton. (1985). A contribution to the population biology of horseshoe crabs, *Limulus polyphemus* (L.), in Delaware Bay. Estuaries 8:363-372.
Swan, Benjie Lynn. (1997). Personal communication on a tag return from an adult female tagged at Reeds Beach, on Delaware Bay, New Jersey on 23 May 1988 and recovered, alive, at Highs Beach, NJ on 31 May 1996. (also in: Swan, Benjie Lynn, William R. Hall Jr. and Carl N. Shuster, Jr. 1998. An Overview of Horseshoe Crab Spawning Activity Along the Shores of Delaware Bay, 1990-1997: 35 pages processed.

Chapter 6

Biomedical Products from the Horseshoe Crab: Lessons for Sound Resource Mangement and Conservation

THOMAS J. NOVITSKY
President/CEO, Associates of Cape Cod, Inc., Cape Cod, Massachusetts

1.   INTRODUCTION

In the 1960's discoveries were made at the Marine Biological Laboratory, Woods Hole, Massachusetts that would lead to the commercialization of a product from the blood cells of the horseshoe crab. This product, *Limulus* amebocyte lysate (LAL), is today a required test on most injectable drugs, invasive medical devices, and vaccines used in North America, Europe, and Japan. Although small, the LAL industry in the United States is regulated by the U.S. Food and Drug Administration. From the outset, the FDA as well as the original LAL manufacturers were concerned about the welfare of the horseshoe crab (Galler, 1979). Unfortunately, current threats to the survival of the horseshoe crab are from outside the industry. Will the LAL industry survive? Most likely it will. We already have genetically engineered replacements being tested in the laboratory. However, the horseshoe crab remains an inestimably valuable resource for new discovery. Already in our laboratory and those of our collaborators we have identified several potentially useful substances with therapeutic/diagnostic applications.

The collection, handling, and live return of horseshoe crabs by the LAL industry will be discussed. The current state and potential of the LAL industry in China and Japan will also be reviewed. New and anticipated discoveries from the horseshoe crab will be highlighted to provide additional reasons for stricter management and conservation of this amazing creature.

Limulus in the Limelight, Edited by John T. Tanacredi
Kluwer Academic/Plenum Publishers, New York, 2001

2.   DISCOVERY AND COMMERCIALIZATION OF LAL

In 1956, Frederick B. Bang, a scientist with Johns Hopkins, doing summer research at the Marine Biological Laboratory (MBL) in Woods Hole, Massachusetts described a disease of the horseshoe crab, *Limulus polyphemus* (Bang, 1956). This discovery lead to a research project, which focused on the phenomenon, related to the infection, namely the action of blood clotting in Limulus. With the collaboration of Jack Levin, the nature of clotting was elucidated and the key role of endotoxin was implicated (Levin, and Bang, 1968). As a result of their work, a reagent, *Limulus* amebocyte lysate (the in vitro clotting system, LAL) was proposed as a test for endotoxin. Further research by Levin and colleagues (Cooper *et al.*, 1970) aroused interest in the clinical applications of LAL as a rapid test for endotoxemia and gram-negative bacteremia.

3.   PHARMACEUTICAL USE AND FDA

Pharmaceutical manufacturers also became interested in the LAL test (Cohen *et al.*, 1979). Their application was as a substitute for the pharmacopeial pyrogen test. Since the most common pharmaceutical pyrogenic contaminant is endotoxin, the LAL test was a logical replacement. The impetus for change was also driven by the fact that the pyrogen test utilized rabbits, was expensive, time consuming and insensitive, while the LAL assay could be performed in any laboratory, was less expensive, quick, and extremely sensitive. With pharmaceutical interest, came United States Food and Drug (FDA) regulation. In 1977, three manufacturers were licensed to produce LAL. These regulations allowed pharmaceuticals to substitute LAL for the pyrogen test. Today, there are five U.S. and one Japanese LAL manufacturers licensed by the FDA. There are also at least 6 manufacturers in China and a number of other manufacturers who produce LAL for "research only" for unregulated pharmaceutical use. Recently China has held a meeting to formulate LAL standards in order to bring their industry up to western norms (Watson, 1979).

4.   SUPPLY OF HORSESHOE CRABS

There are four species of horseshoe crabs worldwide with the abundance of two of the species directly related to the LAL industry. The species that occurs in North America, *Limulus polyphemus,* ranges from Maine to the

Yucatan and gave the LAL test its name. The population center of this species is the Delaware Bay with numbers and size of individuals falling as one goes north or south. It is not surprising therefore that the LAL industry in the U.S. is located between Massachusetts and South Carolina. The species occurring in Japan and China is *Tachypleus tridentatus,* providing for the LAL (sometimes referred to as TAL) industry there. *T. tridentatus* records the largest individual specimens. Today in Japan this species is threatened and considered a national monument. Most probably habitat destruction due to coastal development has led to its decline. Although crabs are still abundant on the Asian mainland, concern is already being raised by China over this resource (Watson, 1979). Elsewhere, there are two additional species, *T. gigas* and *Tachypleus* or *Carcinoscorpius rotundicauda* occurring in India, Thailand, Malaysia, Philippines, and Borneo. *C. rotundicauda,* the smallest of the species, can be found in tidal rivers. Although LAL can be made from these species, their size and abundance limits any significant industry. In Singapore, research with *C. rotundicauda* has demonstrated that some of the enzyme components necessary to produce the LAL reagent can be genetically engineered (Proceedings from the First Chinese Conference on Endotoxin and LAL/TAL Methodology).

In the United States, taking and proper handling of horseshoe crabs for LAL use has been regulated by the FDA since 1977. It is required that all animals be returned to their place of origin immediately following extraction of their blood for LAL production. Although some mortality inevitably occurs during the collection / extraction process, the vast majority survive (Pui *et al.,* 1997). At the Associates of Cape Cod, Inc.'s facility, horseshoe crabs are processed between the months of May and September. Each crab is bled only once during that time. Seasonal repeats are avoided by checking new captures for fresh scars on the dorsal arthroidial membrane, which separates the cephalothorax from the abdomen. It is through this membrane where a large gauge needle is inserted to access the hemolymph. The blue hemocyanin (copper containing oxygen carrier) containing hemolymph contains the amebocytes, which in turn contain the enzymes, which comprise LAL. A typical horseshoe crab will yield between 50 and 100 ml of hemolymph, or about 20- 30% of its hemolymph volume. Although care is taken during the extraction process, the horseshoe crabs circulatory system design prevents exsanguination. It takes a week or two to regain hemolymph volume and several months for the amebocyte count to reach pre-extraction levels, hence the once-a-season extraction rule.

The final LAL reagent is freeze-dried. LAL is used after reconstitution of the dry material with endotoxin-free water or buffer. The test can be run in one of three ways: 1) gel-clot formation: an equal volume of LAL is mixed with sample in a small test tube and incubated at 370°C in a water or dry bath for one hour. Formation of a gel which will withstand 180° inversion is indicative of endotoxin; 2) turbidity formation: Formation of turbidity (read with an optical reader or spectrophotometer) under similar conditions as the gel-clot test is indicative of endotoxin; and 3) chromogen release: Formation of a yellow color as a result of release of paranitroaniline (a chromophore attached to a synthetic substrate added to LAL) under similar conditions as the other methods is indicative of endotoxin.

5.   APPLICATIONS

Although the LAL test is designed only to detect endotoxin, the presence of endotoxin can be indicative of a number of problems from the pharmaceutical industry to the environment. Since endotoxin is unique to gram-negative bacteria, the LAL, under controlled conditions, is also indicative of bacterial presence or contamination. Table 1 outlines the various applications of LAL.

Table 1. Applications of LAL

PHARMACEUTICAL INDUSTRY
Pyrogen Testing
Biologics - Medical Devices - Parenteral Drugs - Artificial Kidneys
Detection of Endotoxin
Cell Culture Media - Raw Materials & Process Water
Products from Genetic Engineering

CLINICAL LABORATORY
Pyrogen Testing of Solution Prepared In-house
Disease Diagnosis
Spinal Meningitis - Septicemia, Bacteremia & Endotoxin
Bacteriuria Gonorrhea - Miscellaneous Diseases by Gram-negative Bacteria

FOOD INDUSTRY
Quality of Fish, Ground Beef & Milk Quality of Frozen Food - Quality of Water

ECOLOGICAL APPLICATIONS
Distribution of Aquatic Bacteria
Monitoring of Stream Pollution
Monitoring Ocean Dumping

6. FUTURE AND MANAGEMENT FOR BIOMEDICAL USE

The LAL industry has always supported strong conservation measures regarding the horseshoe crab. It is ironic that while the FDA requires the live return of horseshoe crabs to the water, local shell-fishing and state regulations often list the animal as a shellfish predator and recommend its stranding above the high water line should it be found or captured. Recent data and resulting draft regulations regarding horseshoe crabs in the center of its range are encouraging. For its part, the LAL industry has always been exploring ways to reduce reagent consumption, and has recently been examining genetically engineered replacements for the test. Furthermore, the LAL industry is small, spread out along the Eastern seaboard and is mature (LAL has been used since 1977). Dramatic increases in LAL consumption are not anticipated.

7. OTHER PRODUCTS FROM THE HORSESHOE CRAB

Although the LAL test is currently the major commercial product from the horseshoe crab, others exist. In Japan, a product which detects glucans (fungal cell components) is sold as a diagnostic test for deep seated mycosis. This test is also based on an extract from the horseshoe crab amebocytes. It is likely that this test will be approved in Europe and the U.S. as there are currently no tests of this type available. Two other products, MicroQuikChek and END-X, made from another horseshoe crab protein, *Limulus* anti endotoxin factor (LALF) or its genetically engineered counterpart, endotoxin neutralizing protein (ENP) are currently available. MicroQuikChek is a rapid (2 minute) test for detecting contamination in metal-working fluids and other liquids. END-X is a research laboratory tool for removing endotoxin contamination from experimental drugs. Although these products can now be made from ENP (produced by yeast in the laboratory), this substance would not have been discovered without the availability of horseshoe crabs for research. A number of other products with industrial, clinical, and environmental uses are also currently under development. One interesting product which could help the horseshoe crab help itself is the attractant which imparts the bait quality to horseshoe crabs when fishing for eels and conch. Should this attractant be found and characterized, a synthetic "bait" should be possible through genetic engineering. Thus, at least one pressure on the crab, as eel and conch bait, could be removed.

8.   CONCLUSION

The story of the horseshoe crab is as marvelous as it is fascinating. Not only is the animal interesting from a biological and natural history standpoint, its usefulness to mankind is commendable. From use of telsons to tip spear points of North American coastal tribes, an unmatched eel and conch bait, to providing safe pharmaceuticals, and diagnosing disease, the horseshoe crab deserves our respect and protection.

REFERENCES

Bang, F.B. (1956) A bacterial disease of Limulus polyphemus. Bull. Johns Hopkins Hosp. 98:325-351.

Cohen, E., Bang, F.B., Levin, J., Marchalonis, J.J., Pistole, T.G., Prendergast, R.A., Shuster, C.S. (1979) Bull. J. Hopkins Press.

Cooper, J.F., Levin, J., and Wagner, H.N., Jr. (1970). New rapid in vitro test for pyrogen in short-lived radiopharmaceuticals. J. Nuc. Med 11:310.

Galler, S.R. (1979) Limulus polyphemus, a target of opportunity. In: Biomedical applications of the Horseshoe Crab (Limulidae, Progress in Clinical and Biological Research, Vol. 29, edited by Cohen, E.New York: Alan R. Liss, Inc., p.107-108.

Levin, J. and Bang, F.B. (1968) Clottable protein in Limulus: its localization and kinetics of its coagulation by endotoxin. Thromb. Diath.Haemorrh. 19:186-197.

Proceedings from the First Chinese Conference on Endotoxin and LAL/TAL Methodology Conference in Ningbo, China. Held at the Zhong Xin Hotel, Ningbo, China, October 29-November 2, 1997.

Pui, A.W.M., Ho, B, and Ding, IL. (1997) Yeast recombinant Factor C from horseshoe crab binds endotoxin and causes bacteriostasis. Journal of Endotoxin Research 4(6):391-400.

Rudloe, A. (1981) The effect of heavy bleeding on mortality of the horseshoe crab, Limulus polyphemus, in the natural environment. Invertebr. Pathol. 42:167-176.

Watson, SW. (1979) Biomedical Applications of the horseshoe crab (Limulidae), Progress in Clinical and Biological Research, Vol. 29, New York: Alan R. Liss, Inc.

Chapter 7

Issues and Approaches in Regulation of the Horseshoe Crab Fishery

JOSH EAGLE
Staff Attorney, National Audubon Society, Washington, D.C.

1. INTRODUCTION

Several years ago, local and national conservation groups began to voice concern over the apparent decline of Horseshoe crabs ("Crabs") and migratory shorebirds on the shores of the Delaware Bay. Several key studies seemed to show that both the crabs and the four species of migratory shorebirds that rely on Crab eggs for fuel in their annual migration were experiencing dramatic population declines.

The relationship between crabs and these four bird species – Red Knots, Ruddy Turnstones, Semi-palmated Sandpipers and Sanderlings – is a unique one. Each spring, major portions of these species' world populations visit Delaware Bay beaches during the spawning ritual of the Crab. Viable crab eggs remain buried in the sand. There are enough eggs left on the beach surface during and after spawning to support birds' during that period and they can increase body weight by as much as 100%. Weight gains of migratory shorebirds during the Delaware Bay stopover are among the highest ever documented in the world. Without enough Horseshoe Crab eggs to eat, it is quite possible that birds would either fail to reach arctic breeding grounds or would arrive with insufficient energy to initiate nesting (Niles and Clark, 1997; Harrington, 1996).

Scientists and conservation groups became concerned after noticing that a number of studies showed significant declines in the numbers of crabs coming onto the beaches to spawn and in the density of Crab eggs on those beaches. (Limuli Labs, Inc., 1997; Botton and Loveland, 1997). A trawl

Limulus in the Limelight, Edited by John T. Tanacredi
Kluwer Academic/Plenum Publishers, New York, 2001

survey conducted in the Delaware Bay by the State of Delaware showed a similar decline, correlating almost exactly with the decline shown in the spawning survey (Delaware 30-Foot Trawl Survey). Aerial counts of shorebirds stopping through the Bay area revealed a steady decrease in populations, from a peak count of 450,000 in 1986 to about 130,000 in 1997 (Niles and Clark, 1997).

What was causing these dramatic changes in the Delaware Bay ecosystem? The most easily identifiable, direct cause was an increase in the take of Horseshoe crabs by commercial fishers. Starting in the late 1980's, eel and conch food markets in the United States, Europe and the Far East created a strong demand for Atlantic coast eel and conch. Horseshoe crabs had been identified by the eel and conch fishers as the most effective bait for use in these fisheries. As a result, both of these fisheries relied almost solely on adult Horseshoe crabs for bait. Due to the increase in demand for bait, numbers of Delaware Bay crabs landed in Maryland, New Jersey and Delaware skyrocketed. For example, National Marine Fisheries Service data show an increase in landings in those three states from 685,648 pounds (170,000 to 230,000 individuals[1]) in 1990 to 3,438,712 pounds (860,000 to 1.3 million individuals) in 1996 (NMFS Horseshoe Crab Landings 1965-1998)[2]

During these years of increased landings, the Horseshoe Crab fisheries in the three states were completely unregulated. Horseshoe crabs were seen as an "unimportant" or "trash" species whose fate was irrelevant. Heightened public awareness of three facts helped to change this view. First, as noted, at least four species of migratory shorebirds depend upon Crab eggs for fuel during their annual migration. Second, the migratory shorebirds provide the foundation for a multi-million dollar ecotourism industry in the Delaware Bay area. Finally, a chemical derived from Crab blood is essential to public health.[3]

As a result of this new awareness, government officials were forced to acknowledge that crabs were not "unimportant". Under pressure from conservation groups, each state developed regulations to control the fishery.

---

[1] Various conversion factors, from 2.7 to 4 pounds per Crab, are used in converting numbers of pounds to numbers of individuals.

[2] The three primary methods of taking Horseshoe crabs are trawling, dredging and hand collection. Trawling and dredging occur mainly in deeper waters, up to 12 miles off the Atlantic coast. Horseshoe crabs are taken by directed fisheries and as by-catch. Smaller numbers are taken in other gear types such as gillnets.

[3] The chemical, Limulus amoebocyte lysate (LAL), is extracted from blood drawn from living crabs, who are later returned to the sea. LAL is used by the pharmaceutical industry to detect the presence of bacteria in injectible drugs and surgical implants, such as prosthetic devices.

These regulations vary significantly from state to state. The purpose of this paper is to:
1. Identify the key provisions in each state's regulations;
2. Identify the kinds of tools each state has chosen to use in attempting to limit the take of crabs;
3. Discuss some political factors involved in the development of the regulations; and
4. Examine the resulting regulations and their potential for effectiveness.

2. DELAWARE

The key provisions of the Delaware regulations (7 Del.C. 1902, 2701) are as follows:
1. No take of crabs is allowed on any state or federal lands during spawning season (May and June), except on the state lands along Port Mahon Road.[4]
2. Take along Port Mahon road is limited to two days per week during the spawning season.
3. Take on private lands is limited to three days per week during the spawning season.
4. The state will only issue up to 50 hand collection permits.[5]
5. The state will only issue up to 6 dredging permits.[6]
6. Dredgers may take up to 1500 crabs per day.
7. Crabs cannot be transported in any vehicle larger than 300 cubic feet.
8. Monthly reports must be filed by all those holding hand collection, dredge or eel permits.

Delaware has taken a multi-faceted approach to reducing the take on crabs. The state has:
1. Limited the times of year, and the days of the week, when crabs may be taken (Nos. 1,2 &3, above).
2. Limited the number of people eligible to take crabs (Nos. 4 and 5).
3. Limited the amount of daily take by those eligible (No. 6).
4. Limited the physical area where crabs may be taken (Nos. 1, 2 and 3).
5. Limited the method of transporting crabs (No. 7).

---

[4] The spawning season, which varies, but typically occurs in May and/or June, is the only time of the year when Horseshoe crabs come onto beaches where they can be taken by hand.
[5] In addition to these individuals, those people holding eel fishing licenses are also entitled to take Horseshoe crabs by hand.
[6] This number is not contained in the regulations, but eligibility requirements limit the field of possible permittees to six.

Although there is no doubt that these are all limitations on the fishery, there is no guarantee that in the end less crabs will be taken with these regulations than without them. Delaware has failed to address the critical variable effort.

From 1993 to 1994, take of crabs in Delaware more than doubled, going from about 30,000 crabs to 75,000 crabs. From 1994 to 1995, take nearly doubled again, and from 1995 to 1996, again. By 1996, 245,889 crabs were taken by hand and by dredge. These numbers seem to indicate that the upper levels of effort have not been reached. In the absence of regulations, landing increases indicate that effort was doubled or nearly doubled every year between 1993 and 1996.

In promulgating the new regulations, the state predicted that take would be reduced by 44% from 1996 levels. These predictions were arrived at simply by looking at the number of days available for take of crabs, which were reduced by about 44% under the new regulations.

However, there is not evidence to indicate that permittees could not simply increase their effort by 44% so that the 1998 landings would be the same as 1996. In fact, because of restrictions taking effect in other states, and the resulting increase in market price for crabs, it is probable that effort will increase by even more, so that 1998 landings will far exceed 1996 landings.[7]

These regulations are thus the perfect example of a political compromise. Elected officials and their agency employees can mollify conservation concerns by pointing to the enactment of "tough" new regulations with a projected 44% reduction in take. On the other hand, they will not face the ire of commercial watermen, who will still be able to take as many crabs as they can sell.

This result in Delaware was the result of several key factors. First, the Governor's office did not become involved in the process of writing the regulations. This task was left to employees of Delaware Department of Natural Resources and Environmental Control (DNREC), who appear to have acted primarily to prevent a political train wreck between conservation and fishing interests. These officials publicly and privately questioned data that showed a significant decline in Crab populations. They were reluctant to harm their relationship with the watermen, with whom they work closely on a wide variety of issues. They expressed a fear that strict cuts would be magnified by the upcoming Atlantic States Marine Fisheries Commission Horseshoe Crab Fishery Management Plan.[8]

---

[7] No data are yet available on 1998 landings.

[8] The ASMFC plan, schedule to be in place by 1999, might for example cut state quota's by a percentage. If 1998 were going to be used as a baseline to determine Delaware's 1999 allocation, it would be in the state's best interest to keep1998 landings as high as possible.

As a result, the Delaware regulations contain several key flaws. First, as mentioned, they fail to address the effort variable. Second, they leave state lands (along Port Mahon Road) open to a natural resource free-for-all. This is uncommon, if not unheard of, in the management of public lands. For example, timber sales from state or federal lands are made based on board feet; there is not open season where timber dredging is still permitted under the Delaware regulations. Dredging describes a process of scraping the ocean floor for harvesting ocean life. Such a practice destroys everything in its path, including plant life and non-targeted species. It is akin to a forest clear-cut, and like a clear-cut, many years are required before ocean floor habitat is regenerated.

The most effective part of the Delaware regulations may turn out to be the restriction on size of transport vehicle. In past years, tractor-trailers would line up along Port Mahon Road, loading up with tens of thousands of crabs. Now, buyers will have to use small pick-up trucks to move the crabs. This may or may not reduce take by making it more time consuming and expensive to transport crabs.

3.   NEW JERSEY

The key provisions of the New Jersey regulations (NJSA 7:25-18.16) are as follows:
1. No crabs can be taken by mechanical means, e.g., dredging or trawling.
2. Hand collection of crabs limited to two days per week from April 1 through August .
3. The state will only issue about 35 hand collection permits.[9]
4. Monthly reports must be filed by all those holding hand collection permits.

Like Delaware, New Jersey has taken a multi-faceted approach to reducing the take of crabs, while choosing not to impose an overall cap on the fishery. The state has:

Limited the method by which crabs can be taken (No.1).
Limited the times of year, and the days of the week, when crabs may be taken (No.2).
Limited the number of people eligible to take crabs (No.3).

Like Delaware, New Jersey has failed to address the effort variable. By banning efficient mechanical means to taking crabs, however, it is less likely that landings will increase from previous years.

---

[9] This number is not contained in the regulations, but eligibility requirements limit the field of possible permittees to about 35.

That is not to say it would be impossible for New Jersey watermen to take the same or greater number of crabs. One effect of the tightening of regulations along the Atlantic seaboard, i.e., decreased supply, is an increase in market price for crabs as bait. Thus, the 35 New Jersey permittees now have incentive to take crabs full-time during the season. This is a change from past practices, where those holding permits were primarily jack-of-all-trade watermen, taking crabs only part-time. Assuming that a permittee can take 2000 crabs per day[10] for sixteen days (two days a week times eight weeks when the crabs are on the beaches), 35 permittees could potentially take 1,120,000 crabs. This would be about two or three times the highest-ever New Jersey annual landings of 1.8 million pounds of crabs (450,000 to 660,000 individuals) in 1996.

The current New Jersey regulations are a product of a 1997 battle between the executive branch and the anachronistic New Jersey Marine Fisheries council, a board of fishing industry representatives that has veto power over fisheries regulations. In New Jersey, unlike in Delaware, the Governor's office took a strong position on the Crab issue, issuing strict regulations on the take of crabs. The Council, using its statutory power, rejected these regulations, which would have put a daily limit of 100 crabs per day on hand collectors. As a result of this limit, together with a ban on mechanical collection, annual take would have been limited to 56,000 individual crabs (35 permittees taking 100 crabs on each of sixteen days).

National Audubon, New Jersey Audubon and the American Littoral Society sued the Council on several grounds, including the fact that it had chosen to ignore scientific evidence in making its decision. The current regulations, which do not include a daily limit on hand collectors, are the result of a settlement to that case that was brokered by the state.

The resulting regulations have similar flaws to those found in the Delaware regulations. As in Delaware, public lands are open to a free-for-all sixteen days per year. This free-for-all jeopardizes the large ecotourism industry in southern New Jersey that is based primarily on watching migratory shorebirds. This industry generates approximately $31 million in Cape May County alone. More important, the regulations leave open the possibility that the ecosystem will be seriously damaged, threatening not only local businesses, but crabs, migratory shorebirds, and other dependent species.

---

[10] This potential daily catch was reported to the author by several watermen and state officials.

4.   MARYLAND

The key provisions of the Maryland regulations (COMAR 08.02.10.01) are as follows:

1. There is a yearly cap on landings of 750,000 pounds of crabs (about 190,000-280,000 crabs).
2. There are daily limits on the number of crabs that can be taken by permittees.
3. The fishery is limited to nine permittees who have shown historic participation in the fishery.
4. There is no take allowed within one mile of the coast.
5. No take is allowed between December and March and only minimal take is allowed during the spawning season (April through July).
6. Reporting is required.

The tools used by Maryland are similar in some ways to those used in the other states, with the significant exception of the yearly quota.

1. There is a yearly cap on landings.
2. The times of year when crabs can be taken are limited.
3. The number of available permits is limited.
4. The amount of daily take is limited.
5. The physical area where crabs can be taken is limited.

There were several key factors in the development of these regulations. As in New Jersey, the Governor of Maryland took a strong personal interest and leadership role in this issue. Together with the Department of Natural Resources (DNR), and important state legislators, the Governor's staff was able to fashion effective regulations in a short period of time. DNR staff were more willing to accept the available scientific evidence showing population declines than their counterparts in other states. This may have been a result of experience with Striped Bass fishery problems (debated intensely before being proven) and pfiesteria outbreaks (showing the delicate state of coastal ecosystems).

As a result, Maryland has the most effective and predicable state regulations on the Horseshoe Crab fishery. The yearly cap provides this predictability in the form of assurances that only a certain number of crabs can be taken. By basing the cap numbers on historical average going back thirteen years, Maryland took a conservative approach, helping to ensure at least stability in Crab populations until the ASMFC plan is completed.

5.   CONCLUSION

It is clear that there are numerous management tools available for regulating the Horseshoe Crab (or any other) fishery. It is equally clear that the only tool that can guarantee a specified objective is a yearly quota. Other tools are susceptible to being rendered ineffective by an increase in effort by fishers.

It is interesting to note that implementation of the most effective regulations was driven by powerful executive leadership. In Maryland, and in New Jersey – prior to the intervention of the Fisheries Council there – the Governors and their staffs were crucial to the process. The reason for this is simple. In the opacity and complexity of data that surrounds every fisheries issue, agency bureaucrats are reluctant to act. This is not a condemnation of these hardworking people. However, fisheries data – especially stock assessments – will always be open to debate. The ocean is a large, dark place, and it can easily be argued (and frequently has been argued by fishing interests) that data collectors are missing information. Thus, important decisions must often be made based on a combination of data and intuition. Such decisions, when they will impact large numbers of people, are most easily made by elected officials.

REFERENCES

Botton and Loveland 1997. Personal communication.
Harrington, B. 1996. The flight of the red knot. W.W. Norton and Co., New York. 192p.
    Limuli Labs, Inc. 1997. 1996 Census Results. Unpubl. Rep. to the Delaware Bay
    Shorebird Working Group. Cape May Courthouse, NJ.
Limuli Labs, Inc. 1997. Personal communication Benji Swan.
Niles, L. and K. Clark. 1997. The decline of the Delaware Bay stopover for migratory
    shorebirds. Unpublished draft.

Chapter 8

The Life History of Horseshoe Crabs

ROBERT E. LOVELAND
Department of Ecology and Evolution, Cook College, Rutgers University

1. INTRODUCTION

Nearly 1,500 miles south of New York City, on the northeast coast of the Yucatan Peninsula, there is a wildlife refuge just outside the little town of Celestun. A causeway to the town traverses a large shallow lagoon, now muddy with calcium carbonate sediments derived from the limestone plate that forms the country rock of the entire peninsula. Large flocks of salmon red flamingos march across the shallows of the refuge, winnowing small crustacea and algae from the rich waters. Huge fresh water springs, called cenotes, pour enormous quantities of fresh water into the lagoon, keeping it brackish. Remarkably, the water source for these springs is in the high eastern slope of the Sierra Madre in Guatemala. Among the spectacular features of this region of Central America is the fact that Celestun lies on the Gulf of Mexico, not far from the impact zone of a huge asteroid that brought an end to the Cretaceous Era over 65 million years ago. Not the least feature is that on the beaches of Celestun can be found the southern-most population of the Atlantic horseshoe crab, *Limulus polyphemus*. These strange Merostomatids cannot be found again until one travels all the way around the Gulf to the panhandle of Florida. No one knows why a small population of horseshoe crabs till exists on the Yucatan Peninsula, so close to the spot where the world's largest explosion set in motion the demise of countless species including the mighty dinosaurs. Horseshoe crabs survived the "nuclear winter" which was created by the impact of the asteroid – most other species did not.

Limulus in the Limelight, Edited by John T. Tanacredi
Kluwer Academic/Plenum Publishers, New York, 2001

2.   SOME HISTORICAL NOTES

Just a few miles south of New York City there is a spit of land jutting several miles into the Atlantic Ocean, forming the border of a shallow bay, and bisecting the Raritan Estuary. Sandy Hook is the name of this remarkably well-preserved and extensive sandy beach. Most of the area is currently part of Gateway National Recreation Area, and is managed by the National Park Service. On a warm summer day, countless tourists wait in line in their cars to gain entrance to the magnificent beaches of the park. Few of these tourists know that just behind them, on the south side of the road, there is a tall promontory composed of sandy sediments, capped with the famous Twin Lighthouses. The sediments that compose the surrounding high hills, called the Atlantic Highlands, were deposited over 65 million years ago – they are part of a ribbon of sediment that stretches from Sandy Hook southwest to Camden on the Delaware River. These sediments contain a rich assemblage of fossils that were formed at the end of the cretaceous, when dinosaurs roamed the beaches of New Jersey. In fact, the first dinosaur ever discovered in the United States came from marl pits in the town of Haddonfield, within the ribbon of these Cretaceous deposits. Even though the asteroid stuck the earth just north of the Yucatan Peninsula, the destruction was felt world side, and many species perished, even along the coast of what is now New Jersey. But not the horseshoe crab. Today, thousands of these helmet shaped creatures emerge onto the sandy banks of Plum Island, in the park, to lay their eggs in the spring. New Jersey remains the epicenter of the only horseshoe crab species in the entire Atlantic Ocean. They are not found in Europe, or Africa, or South America, or anywhere within the entire eastern Pacific basin. *Limulus polyphemus* survived into the Coenozoic (the present), although fossils of its progenitor, *Mesolimulus*, are found in the Cretaceous Solenhofn limestone deposits of Germany. For reasons that are still a mystery, we have no idea where the present day horseshoe crab species came from, or why it is limited in its distribution to the Mid-Atlantic region of eastern United States, with one outlier population in Mexico. Geologists claim that there is a remarkable similarity of the rocks of the Appalachian Mountains to certain areas of South America and Antarctica. Perhaps some ancient island arc climbed its way through the Panama region, drifted to the mid-Atlantic states, and carried with it the predecessors of the Atlantic horseshoe crab from some now extinct southern Pacific form. Even more puzzling is the apparent lack of fossils of *Limulus polyphemus* anywhere – we have simply not found any pat evidence of this species within known fossil assemblages of North America. These marine arthropods are unusual in that they lack calcium carbonate in their exoskeleton. It may be that their chitinous "shell" simply breaks down very rapidly, well before the process of fossilization can occur. Nevertheless, *Limulus polyphemus* is with us today – for how long, is still a mystery.

The oldest reference to horseshoe crabs can be found in the excellent book by Sekiguchi, wherein he cited paintings from the 9th century in China. The early colonists in America called horseshoe crabs "kingcrabs"; even today, one can hear baymen in South Jersey speak of these awkward creatures as kingcrabs (not to be confused with Alaskan kingcrabs, which are true Brachyurans). As was so common in the early zoological literature, horseshoe crabs were taxonomically lumped with mollusks; however, they belong to a group of Arthropods called Chelicerates, to which such animals as spiders, mites, and scorpions belong. Today horseshoe crabs are set apart into a group of Merostomatid chelicerates called Ziphosurans. Paleontologists are not certain when these animals evolved, but Stephen J. Gould has speculated that there may be a merostomatid representative in the Burgess Shale fossils. Strong evidence for early horseshoe crabs, however, does not occur until the late Ordivician or early Silurian, nearly 400 million years ago. By the end of the Paleozoic, there were creatures that were no doubt horseshoe crabs – they are extremely similar to modern day species. This fact is, perhaps, the reason that many popular writers refer to horseshoe crabs as "living fossils" (because of their close resemblance to ancient fossils – strange oxymoron).

Horseshoe crabs also survived the Permian Transition, a period of wholesale extinction throughout the world's fauna. In fact, the best fossil evidence of horseshoe crabs comes from deposits throughout the Mesozioc, when dinosaurs roamed the earth. Today, only four extant species of horseshoe crabs remain – three species in the western Pacific Basin, and one in the mid-Atlantic region of the Untied States.

3. BIOLOGICAL SIGNIFICANCE

The coast of New Jersey in the vicinity of lower Delaware Bay, in the county of Cape May, is the site of the largest population of horseshoe crabs in the world. In the early spring, crabs tend to concentrate on the continental shelf, in a vast region located off the mouth of Delaware Bay, and then slowly move toward the beaches of New Jersey, in the vicinity of Reeds Beach, by Late April or early May. During the first spring tide in May, there is an initial burst of spawning activity, although mated crabs will come ashore on any night after that, especially if there is little wave action. One of the first things you will notice upon observing the masses of spawning crabs in the intertidal zone, is the obvious size difference between the female and the male crabs. A mated pair of crabs will consist of a large female (20-30 cm across the prosoma) with a smaller male (15 – 25 cm across the prosoma) attached to the female. The male hitches onto the female with the use of his second set of appendages, in own as pidpalps; the end of these leg-like structures are modified into claspers which can hold onto the female's opisthosoma so tenaciously that you can literally pick up both animals by the "tail" of the male. The tail is properly known as a telson, and it is not poisonous (as is so often believed by laymen). The phenomenon of attachment is called amplexus, which appears to be a requisite for spawning since rarely would a female approach the beach to lay eggs without her attendant clinging male. In fact, there is accumulated evidence with suggests that a male will amplex with a female long before the now mated pair approach the beach for purposes of spawning. Any female crab may be amplexed by a male of any size, suggesting that there is not size selection by the female for larger males. In fact, mall males from New Hampshire, where there is a relic population of horseshoe crabs that are very small due to geographical isolation, will gladly attach to a giant female from Delaware Bay – and their offspring are viable. In Delaware Bay, most mated pairs will encounter a number of roving suitor (unattached) males as they approach the beach for spawning. It has been shown that while the roving males may participate in the mating process, it is the primary (amplexed) male that contributes the most genetic information to the next generation of crabs. However, the suitor males do contribute to the heterozygosity of the next generation, which is generally thought of as a valuable trait in the reproductive strategy of animals. Within an hour of high tide, most of the spawning horseshoe crabs have moved onto the beach, often in huge numbers – so dense that they form a "pavement" of crabs as far as the eye can see. The female will work her way into the sediments so that she is barely visible. Then she will begin to release eggs, mixing and shaping extruded eggs and sand into a small bolus known as a clutch, which may contain thousands of eggs. At the same time, the male will release sperm that are sucked down to the level of the eggs by ventilatory currents generated by the female. The female will then move forward and deposit yet another clutch of eggs; sometimes as many as 8 clutches can be deposit by a female on a single visit to the beach (females may spawn at least 3 times during the spring). As the tide recedes, the mated pair suddenly turns around, and disappears into the murky waters of the bay. An interesting fact is that the horseshoe crab is the only marine arthropod that has external fertilization. It may be that their reproductive strategy is the reason that the crabs spawn intertidally.

Repeated waves of spawning crabs, tide after tide, will result in some of the eggs moving toward the surface of the beach. Although egg clutches are deposited at least 15 to 20 cm below the sand surface, advancing waves of crabs will disrupt the eggs, and translocate the clutches upward. It is then that huge flocks of migratory shorebirds move to the beaches and consume vast numbers of surface eggs. The annual migration of shorebirds to Delaware Bay is a most astonishing sight to behold. Hundreds of thousands of many species, including red knots, ruddy turnstones, sanderlings, dunlins, and the "peeps" spend their entire daylight time foraging on horseshoe crab eggs. The eggs are quite visible, since they are about 1 mm in diameter and are often bright green; so the birds have no trouble finding the eggs. Many of the shorebirds have flown non-stop from South America all the way to Delaware Bay. So they are exhausted and have lost a lot of body mass by the time they finally alight on the beaches of Delaware Bay. They soon regain their body mass, and "fuel up" for their next venture – to fly to the Arctic for purposes of mating and raising a nest of new fledglings.

Meanwhile, most of the horseshoe crab eggs remain undisturbed within the deeper sediments of the beach; even a sudden summer squall with its accompanying waves will not dislodge the eggs from the sand. Within 10 days to 2 weeks, the eggs will shed their outer layer. It is at this point that one can observe a miniature horseshoe crab developing within the inner transparent membrane of the egg. Eventually, a larval stage appears which is so reminiscent of a trilobite in superficial appearance that they are so named. This first free-living stage has actually molted within the egg membrane several times, but it still lacks a jointed tail. After 3-4 weeks from spawning, the so-called trilobite larva will emerge from its egg membrane, and take up residence between the grains of sand. These larvae may be extremely abundant at times; to observe them, it is necessary only to dig a hole in the middle of the beach at low tide, let it fill with water, and look for the trilobites which will suddenly appear and begin swimming in the water. Around 1990, an unexpected discovery was made in Delaware Bay on a very cold day in January. Beach sediments were brought back to the laboratory and prepared for analysis of organic content. In the preparation of the samples, it became apparent that there were many trilobite larvae in the sand – alive. Systematic sampling of the beaches revealed that horseshoe crab larvae were "resting" at a depth of about 15 cm below the surface; in fact, it was estimated that at least 10% of the initial number of larvae from the previous summer were still resident within the frigid sediments. By April of the following year, all of the trilobite larvae were gone. It has been long known that the trilobites will remain alive at cold temperatures in the laboratory for up to two years, but what were they doing in the sediments in January? Speculation suggests that a fraction of the trilobites "over-winter" in order to hedge their bet against the possibility of depletion by predation. That is, when trilobites emerge form the beach sediments during the summer, many, if not most, are immediately consumed by predators, such as birds, fish, crabs, shrimp, etc. Those that emerge the following April do so when predation pressure is practically non-existent. Incidentally, the trilobite larvae of the Japanese species of horseshoe crab normally over-winters.

We still do not understand the dynamics of movement for trilobite larvae. Clearly, most of these miniature horseshoe crabs remain at or below 15 cm within the beach sediments. While some evidence suggests that larvae can "migrate" to the surface on Florida beaches, this simply does not happen on the beaches of Delaware Bay. It is more likely that heavy surf, which frequently occurs during strong westerly blows on the bay, is responsible for winnowing out the larvae, especially at night. Once dislodged from the waters over the intertidal flats, but not very far since the larvae become positively geotactic and dive for the bottom. Once burrowed into the soft sediments of the flats, the larvae are still vulnerable to crab predation. However, within a day or so, the larvae molt into a stage that is easily recognized as a horseshoe crab with a little telson. This so-called first-tailed stage begins to feed immediately (as far as we can determine, trilobite larvae do not feed; they derive their nutrition from the original yolk of the egg). Within a week or so, the young crab molts again, and begins to move further out onto the flats. By the end of the summer, most of the young crabs will leave the flats – a few will return to the flats the following summer.

For the next 8-9 years, juvenile horseshoe crabs remain in deeper water, molting and growing. There is evidence that a particular size class of crab will all undergo simultaneous molting, because their casts wash up on the beaches, all of uniform size. We know very little about the juvenile phase of the horseshoe crab life cycle. Once the crabs reach a certain size, they undergo a terminal molt and become sexually mature adults. While decidedly larger, the female does not change in appearance at her terminal molt; however, at the smaller male's terminal molt, pedipalps are modified into a pair of claspers that he will use for grasping the female (amplexus). The life span of an average adult horseshoe crab can be as long as an additional 8-10 years, making horseshoe crabs a very long lived marine arthropod, indeed.

Adult horseshoe crabs seem to suffer very little natural mortality. There have been reports of loggerhead turtles and lemon sharks feeding on horseshoe crabs at sea. The greatest natural source of mortality results from the unusual habitat horseshoe crabs have chosen for spawning, namely intertidal beaches. On a calm day, when the crabs swarm onto the beach, most if not all of the crabs will eventually return to the bay. Very few species will attempt to prey on these crabs while they are spawning, except for an occasional Black-backed Gull pecking at them. But when the weather kicks up, and the wind blows, small waves are sufficient to overturn the crab, and they become relatively helpless since it is difficult for them to right themselves. We call the phenomenon of being turned over and helplessly in dire straits "stranding". After a big spawning, during mild surf (the crabs will not even attempt to spawn if the surf is very rough), the beach may become littered with stranded horseshoe crabs. Most stranded crabs will die within the next two tidal cycles if they do not manage to right themselves by using their telson. They can rapidly dry out in the hot baking sun. Large gulls will move in for the kill, feeding on the gills and blood of the crab. It has been estimated that perhaps 10% of the entire spawning population of crabs may be lost to stranding during one summer. Even with this rather staggering loss, the net replacement to the population due to recruitment from previous year's juvenile production will more than offset the loss.

But horseshoe crabs attract human attention. Early in the 20th century, horseshoe crabs were nearly extinct locally. That is because fishermen and farmers cooperated in catching and processing these arthropods in order to manufacture a product that could be used for either organic fertilizer or chicken feed. So many crabs were harvested that the fishery ultimately came to an abrupt end. Then, a remarkable discovery was made at the Woods Hole Marine Biological Laboratory by Frederick Bang. He found that the blood of horseshoe crabs contained a "magic" substance that would bind with bacterial endotoxin. This finding soon became the basis for a diagnostic test to demonstrate that a patient with a high fever or in toxic shock might be infected with gram negative bacterial. Presently, extracted elements or horseshoe crab blood, called *Limulus* Amoebocyte Lysate, or LAL for short, is used in the production of the only accepted material certified by the Food and Drug Administration for testing prosthetic devices, and injectable intravenous fluids, for pyrogenic materials. So horseshoe crabs serve as a living resource for this very valuable material, LAL, since bleeding the crabs does not appear to harm them in any way.

4.   MODERN DAY CONCERNS

Fast forward to the last decade of the 20th century. It was then discovered that horseshoe crabs are extremely useful as bait for catching eels. Individual horseshoe crabs are worth pennies, when they are plentiful, but live eels fetch a handsome fee on the Asian and European markets. Thus began one of the largest fisheries for horseshoe crabs since the latter part of the 19th century. Draggers would routinely catch up to 20,000 crabs per day. Hand-harvesters gathered the crabs from the beaches and shallow waters of the bay. A veritable feast ensued. But then, evidence slowly accrued which suggested that the horseshoe crabs were in troubled waters. There were indications of reduced spawning effort in the Delaware Bay population over the 1990's. The abundance of eggs began to fall, so that by the end of the decade, egg density in the beach sediments dropped by as much as 90% in some areas. Confounding the impact of the horseshoe crab fishery was the obvious loss of habitat all along the bay shore. Over-development of the beaches, bulk-heading, stabilizing beaches with clean fill (another oxymoron), beach buggies, more people utilizing the area, all led to dramatic changes in the appearance of the beaches. Horseshoe crabs are remarkably adaptable creatures, but they cannot, and will not, lay eggs sub-tidally against a bulk-head. It was inevitable, then, that some kind of action to preserve the crabs had to take place. The political force behind this action was largely driven by the age-old relationship between the spawning cycle of the crabs and the utilization of the eggs by migratory shorebirds. A strong dependency between migratory shorebirds and horseshoe crab eggs was eventually demonstrated; the threat to the very existence of the shorebirds was at stake, or so it seemed. But a compromise would be necessary to forestall the clash between conservation of the resource and preserving the traditional lifestyle of the baymen. That compromise was largely introduced by the Atlantic States Marine Fisheries Council, in concert with individual states within the mid-Atlantic zone. At the moment, we are not sure what will be the destiny of the horseshoe crab under the new management strategies. Many researchers, state conservation agencies, fish and game personnel, fisherman, birding groups, and the general public are all cooperating in an effort to discover a strategy, which would lead to a sustainable resource. Clearly, above all, it will be necessary to preserve the horseshoe crab because of the invaluable and pivotal role that it plays in the public health arena.

Because of the concern for what appears to be diminished spawning intensity on the open beaches of Delaware Bay, a concerted effort is being made to assess the role of alternative habitats, which may serve as resources for horseshoe crabs. For example, it was long reported by fishermen that horseshoe crabs migrate into tidal creeks to spawn on the banks of marshes. Such an idea would have previously seemed preposterous, had we not observed this behavior recently. In addition to marshes, horseshoe crabs have been seen to lay eggs within the sediments of offshore sandbars that parallel the beaches, particularly in the vicinity of the delta of a tidal creek. The crabs also appear to use the sandy banks of marsh creeks, especially where creeks empty into the bay. We have even observed horseshoe crabs laying eggs within the sandy soil of an inland forest. This recent spate of anomalous spawning activity may be linked to global ocean rise. That is, as the land subsides and the level of the ocean rises, more and more new habitat becomes available to the migratory crabs. Horseshoe crabs are reluctant to lay eggs on beaches where the veneer of sand is so thin that the eggs are readily dispersed. Indeed, many of the beaches in lower Delaware Bay are now so badly eroded that the depth of the sand is greatly compromised. So horseshoe crabs appear to be searching for areas where there is less erosion, and deeper sand. Some of these marginal habitats have been found to be quite productive of trilobite larvae after the crabs have spawned. Whether the alternative habitats prove to be significant relative to the overall spawning and population dynamics of the crabs is currently under investigation.

5. CONCLUDING STATEMENT

Limulus polyphemus is common to estuaries from Maine to Florida. This species is extremely popular among the public, who spends a lot of time and effort to sojourn to popular spawning grounds to see the crabs in the spring. The linkage of arriving migratory shorebirds to the abundance of eggs continues to fascinate the public. However, there are problems. Reports of fewer crabs along the coast, while often anecdotal are on the increase. Efforts to set aside reserves and conservation areas for the crabs are becoming increasingly negotiated. Even the root causes of global ocean rise are being discussed at the highest level of government, since it is now obvious that the coasts are being inundated. But the horseshoe crab? Well, *Limulus polyphemus* has been around for a long time. It walked among the dinosaurs. It survived the blast of an asteroid that brought an end to the Cretaceous. It adjusted to the last 2 million years of persistent glaciation, with its attendant changes in the configuration of the coastline. It can migrate up rivers and into creeks. It has a great desire to reproduce in large numbers, to lay untold numbers of eggs, although few will survive to become adults. Horseshoe crabs migrate over vast distances in search of new habitat. Their future may appear to be linked to the destiny of humans, but I doubt it. Come bank in another million years – the crabs will no doubt outlast all of us.

REFERENCE

Loveland, R.E., M.L. Botton and C.N. Shuster, Jr. (1996). Life history of the American Horseshoe Crab (*Limulus polyphemus L.*) in Delaware Bay and Its Importance as a Commercial Resource.

Chapter 9

Horseshoe Crab Mangement and Resource Monitoring in New Jersey 1993-1998

PETER J. HIMCHAK and SHERRY L. HARTLEY
New Jersey Division of Fish, Game and Wildlife, Trenton, New Jersey

1. HORSESHOE CRAB RULE AND SUBSEQUENT AMENDMENTS

In response to escalating confrontations between shorebird conservationists and horseshoe crab harvesters on Delaware Bay beaches, the New Jersey Division of Fish, Game and Wildlife Division developed and implemented Horseshoe Crab Rule N.J.A.C. 7:25-18.16 effective May 3,1993. The rule, which minimizes the disturbance of migratory shorebirds while they are feeding on Delaware Bay beaches in New Jersey, also established the Horseshoe Crab Permit, whereby any person harvesting horseshoe crabs is required to possess a permit and to provide monthly reports that include the number of horseshoe crabs harvested, the area of collection, the gear utilized, and other information. The Horseshoe Crab Rule did not restrict the number of horseshoe crabs harvested, but it did restrict times from May 1 through June 7 when horseshoe crab harvesting could take place along the Delaware Bay shorelines. Harvesting was allowed on Monday, Wednesday, and Friday during nighttime hours only.

Although the horseshoe crab harvesting season was under way when the permit system became effective, 141 permits were issued in 1993 (Table 1), and the reported harvest was 281,135 horseshoe crabs. Only 28.4% of those who obtained permits complied with the reporting requirements, but many permits were obtained by individuals who subsequently did not harvest

Limulus in the Limelight, Edited by John T. Tanacredi
Kluwer Academic/Plenum Publishers, New York, 2001

horseshoe crabs. The Horseshoe Crab Permit is free, but both residents and non-residents are charged a two dollar administrative fee to obtain the permit.

The 1994 and 1995 horseshoe crab commercial fishery was conducted under the provisions of the original Horseshoe Crab Rule. Commercial effort did not increase substantially in 1994 when 156 permits were issued (Table 1). The reported total harvest for 1994 was 357,438 horseshoe crabs (Table 2). The two predominant horseshoe crab fisheries were hand harvesting, primarily on Delaware Bay beaches in the springtime and an offshore trawl fishery during the late summer and throughout the fall. Hand harvesting accounted for approximately 40% of the reported harvest and the trawl fishery accounted for nearly 60% of the total reported harvest. Other/unspecified entries within horseshoe crab harvest tables result from incomplete record keeping by some permittees. Though compliance to the reporting requirements of the horseshoe crab permit system had improved, the reported harvest for the years 1993-1995 may be an underestimate of the commercial horseshoe crab landings due to non-reporting and possibly underreporting by some permittees.

TABLE 1			
Horseshoe Crab Permits Issued, 1993 - 1998			
Year	Residents	Non-Residents	Total
1993	110	31	141
1994	132	24	156
1995	291	14	305
1996	342	5	347
1997	330	2	332
1998	39	0	39

TABLE 2							
1994 Horseshoe Crab Harvest							
	HAND		TRAWL		OTHER/UNSPECIFIED		TOTAL	
	NO.	%	NO.	%	NO.	%	NO.	%
DELAWARE BAY	63,108	####	88,752	####	4,855	1.4	######	43.8
ATLANTIC COAST	963	0.3	######	####	-	-	######	34.8
UNSPECIFIED	75,890	####	-	-	320	0.1	76,210	21.3
TOTAL	#######	####	######	####	5,175	1.4	######	#####

In 1994, non-residents harvested 11.6% of the total reported harvest; all were harvested by hand on Delaware Bay beaches. Non-resident permits were obtained by groups of individuals from Virginia and Massachusetts for bait in their eel and conch fisheries back home.

The number of horseshoe crab permits issued in 1995 (Table 1) increased substantially to 305 as the public became aware of an amendment to the Horseshoe Crab Rule that was being developed which might restrict entry into the horseshoe crab commercial fishery. Many permits were obtained very late in the year. Effort in hand harvesting had not increased substantially earlier in the year and effort in the trawl fishery has remained relatively stable throughout all the years of the horseshoe crab permit system. In 1995, the reported harvest was 394,619 horseshoe crabs, again largely attributed to a springtime hand harvest and a trawl fishery later in the year. There was no harvest reported by non-residents, the two major groups of non-residents from Virginia and Massachusetts who had permits in previous years were able to obtain their horseshoe crabs in their own states without harvesting in New Jersey.

Of particular concern to the Division in the horseshoe crab commercial fishery was the growing practice of scraping or "dip-netting" for horseshoe crabs by fishermen in boats drifting along the spawning beaches. The new harvesting technique, though extremely laborious was quite successful since the horseshoe crabs were so heavily concentrated along the beaches during the spawning season.

2. 1996 AMENDMENT TO N.J.A.C. 7:25-18.16

In response to increased concern over the biological condition of the horseshoe crab resource, an amendment to N.J.A.C. 7:25-18.16 was developed in 1996 to provide added protection to spawning horseshoe crabs and to reduce the disturbance to migratory shorebirds feeding on the Delaware Bay waterfront beaches. The amendment was developed by the New Jersey Marine Fisheries Council in coordination with the Endangered and Non-Game Species Advisory Committee, bird conservationists and commercial fishermen.

Adoption of the amendment to the horseshoe crab rule prohibited the harvest of horseshoe crabs on the Delaware Bay waterfront at any time. Hand harvesting would be permitted only in back bays and tidal creeks of New Jersey (minimum of 1,000 ft. from bayfront) on Tuesdays and Thursdays. The amendment did not restrict the use of legal gear (net or other means used) in the prescribed manner in the bay or Atlantic coastal waters except from May 1 through May 31 when all horseshoe crab harvesting was limited to two days a week by hand only. This provision would provide protection to crabs during the peak spawning period. The amendment further authorized the Division to suspend or revoke a horseshoe crab permit for failure to comply with the specified reporting requirements.

In 1996, 347 horseshoe crab permits were issued (Table 1). The reported harvest for 1996 increased substantially to 606,583 horseshoe crabs (Table 4). The proportion of hand harvesting and trawling components in the reported harvest remained relatively constant, since the increased other/unspecified component of the harvest was largely attributed to poor record keeping by many new entrants who were hand harvesting horseshoe crabs.

TABLE 3							
1995 Horseshoe Crab Harvest							
	HAND		**TRAWL**		**OTHER/UNSP ECIFIED**		**TOTAL**	
	NO.	%	NO.	%	NO.	%	NO.	%
DELAWARE BAY	83,444	####	-	-	43,034	####	######	32.1
ATLANTIC COAST	4,413	1.1	######	####	25,881	6.6	######	67.8
UNSPECIFIED	525	0.1	-	-	-	-	525	0.1
TOTAL	88,382	####	######	####	68,915	####	######	#####

TABLE 4							
1996 Horseshoe Crab Harvest							
	HAND		**TRAWL**		**OTHER/UNSP ECIFIED**		**TOTAL**	
	NO.	%	NO.	%	NO.	%	NO.	%
DELAWARE BAY	#######	####	82,390	####	11,625	1.9	######	38.7
ATLANTIC COAST	10,564	1.7	######	####	20,480	3.4	######	46.9
UNSPECIFIED	7,162	1.2	1,000	####	78,975	####	87,137	14.4
TOTAL	#######	####	######	####	######	####	######	#####

Many conservation groups became alarmed over the substantial increase in reported horseshoe crab landings in 1996. They believed the restrictions in harvesting horseshoe crabs resulting from the adoption of the 1996 amendment to the horseshoe crab rule would reverse or level off the increasing horseshoe crab reported harvest. Better compliance to the reporting requirements of the horseshoe crab permit system and the resulting more accurate landings in the commercial fishery, however, may have provided in 1996 the best estimate to date of the magnitude of the horseshoe crab commercial fishery in New Jersey. The suspension of horseshoe crab harvesting privileges to permittees who did not file the required harvest reports in 1996 resulted in many letters from permittees who received suspension notices. The majority admitted that their failure to file harvest reports was due to their belief that if no harvest was taken, no report was necessary. This correspondence confirmed our confidence in the 1996 harvest estimate.

3.   1997 AMENDMENT TO N.J.A.C. 7:25-18.16

The number of permits issued in 1997, 332 (Table 1), declined slightly from earlier years. The declining trend in the number of non-resident permittees over the years is evident (Table 1); it was no longer economical to harvest horseshoe crabs in New Jersey because of temporal and spatial restrictions. Anecdotal information suggests that the majority of horseshoe crabs taken in New Jersey were now being exported to supply bait for other states' eel and conch fisheries.

Following the reporting of some modest late winter/early spring trawl landings, the reported hand harvesting of horseshoe crabs in 1997 reached nearly 260,000 mid-way through the hand harvesting season, surpassing the reported hand harvesting for the entire 1996 season. Data on the monitoring of horseshoe crab egg counts in the sediments of spawning beaches showed dramatic declines in both 1996 and 1997 from an earlier study in 1990. The expanding harvest and declining horseshoe crab egg counts on spawning beaches heightened concern for the conservation of horseshoe crabs and migratory shorebirds. This led Governor Christine Todd Whitman to declare a moratorium on the harvesting of horseshoe crabs in New Jersey. On May 30, 1997, the Department of Environmental Protection (Department) adopted emergency amendments to the rules governing New Jersey's horseshoe crab fishery which imposed a ban on all harvest of horseshoe crabs by hand, dredging, or trawling from the beaches, shoreline, and marine waters of the State. The 60 day period of the ban imposed under the emergency rule expired on July 29, 1997. On that date, the Department adopted on an emergency basis, amendments to N.J.A.C. 7:25-18.16 that continued the ban on fishing for horseshoe crabs with gear (dredge, trawl, or any implement) but allow hand harvesting under limited circumstances.

The 1998 horseshoe crab commercial fishery was governed by the emergency amendment and concurrent proposed amendment adopted on September 25, 1997 which allowed for the harvest of horseshoe crabs by hand collection only (prohibiting all other gear types), from May 1 through June 30, and established criteria for individuals to qualify for the horseshoe crab permit which is necessary to harvest horseshoe crabs.

For 1998, horseshoe crab permits would be available only to those individuals in the eel and conch pot fisheries who are recent participants in the horseshoe crab fishery. The possession of a valid New Jersey miniature fluke or lobster or fish pot license and the possession of a valid New Jersey horseshoe crab permit and reported landings of horseshoe crabs as verified by the Department in each of two calendar years during the period of January 1, 1993 through May 29, 1997 became the criteria used to establish eligibility for individuals to obtain a horseshoe crab permit.

Surprisingly, only 39 applicants qualified for and received a horseshoe crab permit in 1998 (Table 1). Of the nearly 100 applicants who applied but did not quality for a horseshoe crab permit, approximately 89% did not satisfy the qualifying criterion of reporting a horseshoe crab harvest in two separate years. Many commercial eelers and conch pot fishermen who acquired horseshoe crab permits in previous years did not harvest horseshoe crabs for their own bait needs. Horseshoe crabs had been very inexpensive and many eelers and conchers had purchased them for bait from other permittees.

TABLE 5							
1997 Horseshoe Crab Harvest							
(Jan. - May 29)	**HAND**		**TRAWL**		**OTHER/UNSPECIFIED**		**TOTAL**	
	NO.	%	NO.	%	NO.	%	NO.	%
DELAWARE BAY	#######	####	50,958	####	73,692	####	######	88.7
ATLANTIC COAST	6,414	1.6	17,545	4.3	300	####	24,259	6.0
UNSPECIFIED	19,298	4.8	329	####	1,631	0.4	21,258	5.3
TOTAL	#######	####	68,832	####	75,623	####	######	#####

TABLE 6	
1998 Horseshoe Crab Harvest	
(April - August)	**HAND**
	NO.	%
DELAWARE BAY	#######	96.1
ATLANTIC COAST	9,335	3.9
TOTAL	#######	100

4.   1998 AMENDMENT TO N.J.A.C. 7:25-18.16

Prior to the beginning of the horseshoe crab harvesting season, a 1998 amendment to the horseshoe crab rule was adopted allowing for permit transferability among immediate family members and changing the hand harvesting season to April 1 through August 15.

The 1998 horseshoe crab reported harvest was 241,456 horseshoe crabs (Table 6), 96.1% were taken throughout Delaware Bay and 3.9% were taken along the Atlantic Coast. All landings were from hand harvesters in accordance to the provisions of the 1997 amendment to the horseshoe crab rule. Compliance to the reporting requirement of the horseshoe crab rule was excellent.

5.   FISHERY INDEPENDENT MONITORING

The New Jersey Ocean Stock Assessment Program is a coastal trawl survey designed to develop comprehensive baseline data for recreational fishes and their forage items. The survey area consists of New Jersey coastal waters from the entrance to New York Harbor south to the entrance to Delaware Bay (Fig. 1). Samples are collected by trawls of 20 minutes tow duration. Trawl specifications include an 82-ft. (25.0 in) headrope, and a 100 ft. (30.5 in) footrope covered with 2-3/8 in. (6.0 cm) rubber cookies. Five sampling surveys are completed each year (January, April, June, August, and October). Sampling stations for each survey are selected randomly for each stratum (Byrne, 1997).

The New Jersey ocean trawl survey provides a relatively long-term (1988-1997), continuous resource monitoring database (Table 7, Fig. 2), which was being used to estimate the relative abundance of horseshoe crabs in New Jersey coastal waters.

TABLE 7						
New Jersey Horseshoe Crab						
Trawl Survey Data, 1988 - 1998						
	No. of	No. of	Mean Catch per	Standard	Geometric Mean Catch	Standard
Year	Crabs	Tows	Unit Effort	Deviation	per Unit Effort	Deviation
1988	518	65	8.0	30.0	2.2	1.1
1989	1,562	191	8.2	29.9	1.3	1.3
1990	2,823	163	17.3	63.7	1.8	1.5
1991	1,150	188	6.1	20.4	1.3	1.2
1992	1,316	193	6.8	17.2	1.6	1.3
1993	1,014	189	5.4	16.4	1.3	1.2
1994	816	185	4.4	13.8	0.9	1.1
1995	918	187	4.9	15.2	1.2	1.1
1996	1,423	193	7.4	19.8	1.7	1.3
1997	2,021	189	10.7	21.3	2.8	1.4

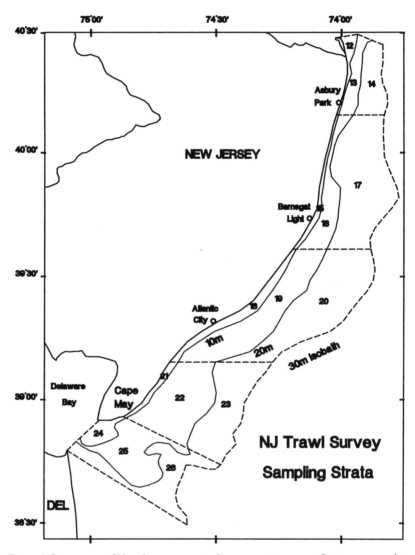

Figure 1. Survey area of New Jersey ocean stock assessment program. Strata corresponds to those of the National Marine Fisheries Service's spring and fall groundfish surveys except at Sandy Hook and Cape Henlopen, where they are truncated. Longitudinal boundaries approximate 9.1 m, 18.3 m, 27.4 m.

The trawl survey catch per unit effort (cpue) was initially calculated with all samples from each year pooled. The annual cpue, calculated as either an arithmetic mean or a geometric mean (Fig. 2) showed high variances around each annual mean value. Horseshoe crab catches were shown to be highly variable, influenced by season and sampling stratum.

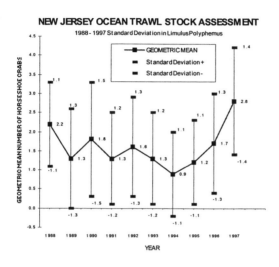

YEAR						
	GEOMETRIC MEAN	Standard Deviation +	Standard Deviation -		Actual Standard Deviation -,+	
1988	2.2	3.3	1.1		1.1	
1989	1.3	2.6	0.0		1.3	
1990	1.8	3.3	0.3		1.5	
1991	1.3	2.5	0.1		1.2	
1992	1.6	2.9	0.3		1.3	
1993	1.3	2.5	0.1		1.2	
1994	0.9	2.0	-0.2		1.1	
1995	1.2	2.3	0.1		1.1	
1996	1.7	3.0	0.4		1.3	
1997	2.8	4.2	1.4		1.4	

Figure 2. New Jersey ocean trawl stock assessment

The appropriateness of trawl surveys designed primarily to catch groundfish in providing adequate data to estimate horseshoe crab cpue from year to year was much debated given the relative inefficiency of trawls to catch horseshoe crabs. In any year, the number of zero horseshoe crab catches far-exceeded catches where horseshoe crabs were caught. Nonetheless, coastal trawl surveys by states and the National Marine Fisheries Service (NMFS) represented the best available long-term information on horseshoe crab relative abundance. The trawl surveys were well designed with random stratified sampling, and sampling methodology within each survey was consistent throughout the time series of most trawl surveys.

In its report (ASMFC, 1998a) to the ASMFC Horseshoe Crab Technical Committee (TC), the Horseshoe Crab Stock Assessment Subcommittee (SASC) recommended the need to "evaluate trawl surveys for their sensitivity and reliability in this stock assessment". The TC recommended that the horseshoe crab stock assessment be considered as a high priority for peer review by the ASMFC.

A Horseshoe Crab Peer Review Panel ("Panel") was convened September 29-October 1, 1998 to review the SASC's horseshoe crab stock assessment report. On the utility of existing trawl surveys to assess horseshoe crab trends in relative abundance, the Panel reported:

> "The Panel reviewed the various trawl surveys and associated analyses used in the Stock Assessment Report and found that they were of little to no value in assessing the status of the horseshoe crabs. This is primarily due to the fact that the trawl surveys were multi-species finfish surveys that were not designed for horseshoe crabs. Horseshoe crabs were viewed as bycatch in these surveys as they were not a targeted species. The gear in many of these trawl surveys precluded the effective capture of the majority of horseshoe crabs that could have been available for capture. For this reason captures were infrequent and produced uninformative data."   ASMFC, 1998b.

The large number of zero catches in the New Jersey trawl survey (Table 8) and the large variance around annual mean cpue estimates (Table 7, Fig. 2) confirm that the catches of horseshoe crabs occur infrequently and when caught, the numbers taken are usually relatively low (i.e., 1 to 10 horseshoe crabs).

In an attempt to reduce the variance of annual cpue estimates, the Panel recommended a post stratification of data by depth stratum since previous data analysis demonstrated a gradiant of horseshoe crab catches from nearshore to offshore. Annual horseshoe crab catch data were separated into inshore (0-5 fathoms), midshore (5- 10 fathoms) and offshore (10- 15 fathoms) strata (Fig. 1). The catch frequency data by stratum showed the same pattern on the infrequency of horseshoe crab catches and when caught, horseshoe crabs were taken in very low numbers (Table 8). The poststratification analysis did demonstrate by comparing annual cpue's with depth stratum that the concentration of horseshoe crabs was greatest near shore in shallower waters and decreased with distance from shore (Fig. 3).

TABLE 8

1988 -1997 NEW JERSEY OCEAN TRAWL STOCK ASSESSMENT OF STANDARD VALUES OF *LIMULUS POLYPHEMUS*

NUMBER OF TOWS PER YEAR

No. HC Caught	1988 MI ND	1988 OF F	1989 MI ND	1989 OF F	1990 MI ND	1990 OF F	1991 MI ND	1991 OF F	1992 MI ND	1992 OF F	1993 MI ND	1993 OF F	1994 MI ND	1994 OF F	1995 MI ND	1995 OF F	1996 MI ND	1996 OF F	1997 MI ND	1997 OF F	TOTAL NUMBER OF TOWS PER YEAR 88	89	90	91	92	93	94	95	96	97	TOTAL AREA 88	89	90	91	92	93	94	95	96	97
Zero	4	5	8	37	36	33	47	34	42	37	46	45	47	50	34	39	27	1	10	10	10	11	10	10	11	7	10	77	62	62	20	18	16	18	18	19	18
1 to 10	1	11	15	28	14	17	13	25	13	26	15	18	14	9	20	23	29	4	7	61	64	64	4	12	6	10	3	0	9	7	8	9	3	5	5	7	3	9
11 to 20	6	7	3	6	2	4	2	7	3	4	6	9	14	9	18	2	5	16	11	10	9	2	3																	
21 to 30	2			3		5						2	5	1	1	2	3	3	4																					
31 to 40			2	1	3		3	2	3	1	2	1		4		1																								
41 to 50			1	1			1	1	1			1					3	2	2																					
51 to 60			1				1	1						2		2																								
61 to 70																	1																							
71 to 80					2	2											2																							
81 to 90																																								
91 to 100																																								
103																	1																							
104																																								
107						1						1																												
112																																								
117																																								
124																																								
128				1										1																										
133																																								
135																	1																							
143																																								
154																																								
159																1																								
163																	1																							
167			1															1																						
194																																								
204		1																																						
215					1													1																						
219			1	1																																				
234			1	1		1		1																																
258																		1																						
288																		1																						
410		1													1			1																						
522																		1																						
Total Tow Effort	2	2	5	74	58	54	66	69	67	66	63	66	67	63	66	65	67	63	6	19	18	18	18	19	16	5	6	6	6
# of Cruises	3	2	6	6	5	5	5	5	5	5	5	5	5	5	5	5	5	5	5	5	5	5	5	5	5	5	2		

In October 1998, the ASMFC adopted the Interstate Fishery Management Plan for Horseshoe Crab, (ASMFC, 1998c). Among many monitoring components required by states is the continuation of trawl surveys to collect data on horseshoe crab weight, number, and prosomal width by sex of individuals collected. It is further recommended that a coast-wide benthic sampling program designed to efficiently capture horseshoe crabs be implemented for future stock assessment in detecting trends in horseshoe crabs relative abundance from year to year.

year	Inshore (0 - 5 Fathoms)	Midshore (5 - 10 Fathoms)	Offshore (10 - 15 Fathoms)
1988	3.0	2.7	1.4
1989	3.2	1.3	0.4
1990	4.7	1.2	0.8
1991	2.7	1.3	0.5
1992	2.8	1.6	0.9
1993	2.7	1.1	0.5
1994	2.3	0.6	0.4
1995	2.7	1.0	0.5
1996	3.9	1.6	0.6
1997	8.6	2.1	1.0

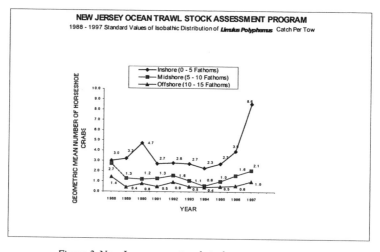

Figure 3. New Jersey ocean trawl stock assessment program

REFERENCES

Atlantic States Marine Fisheries Commission. 1998a. Status of the Horseshoe Crab *(Limulus polyphemus)* Population of the Atlantic Coast. Report to the ASMFC horseshoe Crab Technical Committee. Wash., D.C., 48 pp.

Atlantic States Marine Fisheries Commission. 1998b. Horseshoe Crab Peer Review. Wash., D.C., 12 pp.

Atlantic States Marine Fisheries Commission. 1998c. Interstate Fishery Management Plan for Horseshoe Crab. Fishery Management Report No. 32 of the ASMFC. Wash., D.C., 58 pp.

Byrne, Donald M. 1997. Inventory of New Jersey's Coastal Waters. Final Report, Project *F-15-R-38*. NJ Div. Fish, Game & Wildlife, Trenton, NJ. 18 pp.

Himchak, Peter J. 1996a. I-Horseshoe Crab Management in New Jersey, pages 33-34 In Proceedings of the I-Horseshoe Crab Forum, Status of the Resources, February 23, 1996, Univ. of DE, Sea Grant College Program, DEL-SG-05-97. 60 pp.

Himchak, Peter J. 1996 b. Resource Monitoring of Horseshoe Crabs in New Jersey. Pages 3 1-32 In Proceedings of the Horseshoe Crab Forum, Status of the Resource, February 23, 1996, Univ. of DE, Sea Grant College Program, DEL-SG-05-07. 60 pp.

Steel, R.G.D. and J.H. Torrie. 1960. Principles and Procedures of Statistics. McGraw Hill, New York. 13-20 pp.

Zar, Jerrold H. 1984. Biostatistical Analysis. 2nd ed., Prentice Hall, New Jersey. 18-31 pp. and 484-485 pp.

PART IV

Present Day Investigations

Photo by Don Riepe.

"I am convinced that a quasi-religious movement, one concerned with the need to change the values that now govern much of human activity, is essential to the persistence of our civilization. But agreeing that science, even the science of ecology, cannot answer all questions—that there are other ways of knowing—does not diminish the absolutely critical role that good science must play if our over-extended civilization is to save itself".

Paul Ehrlich, in *The Machinery of Nature.*

Chapter 10

Horseshoe Crab Surveys Using Underwater Videography

CHRISTINE KURTZKE
Fisheries Biologist, Gateway National Recreation Area, Division of Natural Resources, 210 New York Avenue, Staten Island, New York 10305, USA

1. INTRODUCTION

The American Horseshoe Crabs (*Limulus polyphemous*), the only North American species of four Xiphosurans left (Loveland *et al.*, 1996), is estimated to be at least 500 million years old. The oldest fossil forms of Horseshoe Crabs fragments were found 550 million years ago and discovered in early Cambrian rocks. About 300 million years ago, during the Carboniferous Period, and as early as any known dinosaur fossil, the horseshoe crab, whose shape and form we know today, appeared. In fact, these fossil forms are said to be physically similar to the ones I have observed in Jamaica Bay. Both males and females come to the sandy channels and bays of the east coast to breed. The greatest concentration of spawning Horseshoe Crabs are believed to be in Delaware Bay. Approximately 90% of all *Limulus polyphemous* individuals are found between Virginia and New Jersey (Botton and Ropes, 1987), but can also be found along the shores from Maine to the Yucatan Peninsula, Mexico.

In their quest to breed, thousands of Horseshoe Crabs move in the direction of sandy beaches during the full and new moons in April and May (Barlow *et al.*, 1986). Females leave the water and burrow in the sand where they lay up to 80,000 eggs per season. Males, who are frequently attached to the carapace of the female by a pair of specialized front claws, fertilize the eggs she lays by secreting a jelly substance containing sperm (Shuster and Botton, 1985). Waves then complete the process of combining sperm and egg by washing sand over the nest and eggs. The eggs remain buried for

Limulus in the Limelight, Edited by John T. Tanacredi
Kluwer Academic/Plenum Publishers, New York, 2001

approximately 30 days, after which juveniles hatch out and reenter the ocean.

Although given the name Horseshoe Crab in the 1800's, and also known as the 'king crab' and 'pan crab' to fishermen, the Horseshoe Crab is not a true crab at all. It is a marine arthropod of the order Xiphosura in the class Merostomata. Horseshoe crabs are more closely related to spiders and scorpions. However, contrary to these class members' reputations, horseshoe crabs are not predatory; and their only known predator is man. Horseshoe Crab numbers have dwindled over the years due to harvesting for use as fertilizer, eel, catfish and conch bait. They are also important to research and pharmaceutical testing because their blood naturally coagulates in the presence of endotoxins. Formed as byproduct of Horseshoe Crab blood, *Limulus* Amoebocyte Lysate (LAL) is used to test the sterility of injectables, i.e. needles, vaccines, and also heart valves and dental instruments (Cohen *et al.*, 1978). Horseshoe Crabs' lateral eyes also contribute to human ocular research, because they possess long optic nerves, facilitating laboratory work.

In actuality, horseshoe crab populations are declining throughout the range due to overharvesting. Annual horseshoe crab spawning data reveal that from 1991 to 1994, a decrease from 1,225,000 to 535,000 individuals was observed (University of Delaware, 1996). Because of its critical concern status, the first Fishery Management Plan for Horseshoe Crabs Conservation was drafted in October 1998. It is applicable only to Delaware, Maryland, New Jersey and Virginia (Peter Himchak, personal communication). Unfortunately, horseshoe crabs are easily harvested during their spawning season, because they move toward sandy beaches in intertidal zones, and can be caught with a minimal amount of effort and expense. Also critical spawning areas that have been, or will be affected by coastal development and pollution, alter the existing numbers of horseshoe crabs in Jamaica Bay. Bait fishermen, who generally only select egg-bearing females in their traps, also need to be closely monitored or even restricted in their catches. In addition, heavy metals and the effects of these substances on horseshoe crabs influence their survival rate. For example, in a study conducted on juvenile horseshoe crabs exposed to metal laden sediment, it was observed that the chemicals caused mutated juvenile body parts. It was theorized that these mutated parts would negatively affect the horseshoe crabs' reproductive success and survival rates as they matured (Itow *et al.*, 1998). This has already been seen in animals such as alligators, and in amphipods (Borowsky *et al.*, 1993). Clearly, the behavior, ecology and reproductive success of horseshoe crabs needs to be studied further, by both inventory and monitoring procedures, and by observation of them in their natural habitat. This will assist in helping to preserve their habitat.

2. METHODOLOGY

Gateway National Recreation Area, where Horseshoe Crabs enter by the thousands every spring, is an urban park consisting of 26,000 acres; the majority of which is estuarine. Gateways' boundaries include Jamaica Bay in Brooklyn, New York, Great Kills in Staten Island, New York and Sandy Hook in Sandy Hook, New Jersey (Fig. 1). In early spring, *Limulus* enter Jamaica Bay, Crooke's Point Bay in Great Kills, and Spermicetti Cove and Horseshoe Cove in Sandy Hook, from the ocean. On new and full moons from May through July, they gather en masse to lay and fertilize their eggs on sandy beaches. Upon hatching, the juveniles inhabit nearby intertidal flats and shallow waters, which are located in the Cross Bay waters in Jamaica Bay (Fig. 1). As they grow and mature, they move further away from their birth area, migrating into deeper water. Eventually they winter on or near the continental shelf, at depths of 200 to 300 feet, where they will remain until returning again next year.

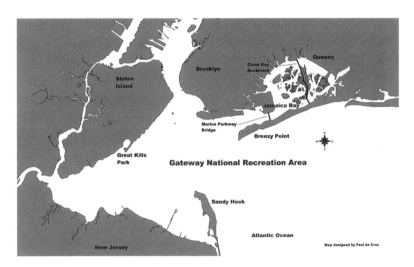

Figure 1. The boundaries of Gateway National Recreation Area showing Jamaica Bay and Sandy Hook.

One of the research projects we attempted was documenting Horseshoe Crab behavior and movement in Jamaica Bay, by using SCUBA Diving and underwater Videography. With the innovative development of the aqualung by Jacques Costeau and Emile Gagnan in the 1940's, prolonged and unencumbered encounters underwater can be accomplished (Sullivan, 1964). Marine researchers in the past have used scuba diving technology for direct observation of, and complete access to, underwater organisms and their habitats. The US Navy MK-5 Hardhat Diving System, which was used consistently for over 50 years (Tanacredi, 1998), led the way for new advances in hardware and technology following World War II. Modern day SCUBA divers still use the Navy dive tables to monitor safe diving times and depths to prevent the 'bends'. The bends is an excess of nitrogen that builds up in tissues. This happens due to either excess time spent underwater at depth, or due to "bounce diving", which is going deep and shallow frequently during the dive. It can cause fatigue and pain, but its most terrifying effect is paralysis, and even death. Since this type of diving causes exposure to increased pressure on divers by the water they are in, scuba diving is also known as "ambient-pressure diving". Constraints that are invariably due to this type of diving include depth and time one can spend underwater safely on a given dive, visibility of the water column, water current effects on position, water temperature effects on the body, and air and climatic conditions in which one is able to dive.

The National Park Service (NPS) has the oldest non-military diving program in the government. About 150 park rangers have diving credentials and dive as part of their job, and the NPS Park Police dive for security and rescue purposes (Lenihand and Brooks, 1997). The reason we use SCUBA diving in scientific research of Horseshoe Crabs is because most of their observed behavior to date has been done out of the water, while breeding on beaches (Shuster, personal communication). To observe and document their underwater behavior, i.e. feeding, burrowing, resting, and swimming, we can get a better idea of their morphometry. In addition, as part of our research, we used underwater photography to aid in our observation, and also to allow experts in the field to observe the horseshoe crab behavior as well. Many marine biologists say that to preserve our ocean species for the ecology and economy harvesting that will be done in the future, disciplines such as biotechnology, ocean engineering, marine biology and marine archeology will be desperately needed. At the Marine Resources Research Institute in Charleston, South Carolina, as in many other Universities and marine research laboratories around the world, non-removal diver census of fishes, and visual census methods (remote video) have been used to access fish stocks.

3.   RESULTS

Members of the Gateway Park Police SCUBA Unit and I dove in Jamaica Bay during the spring and summer of 1999, equipped with an underwater video camera to document Horseshoe Crab behavior. Since the equipment we use is fairly portable, our air tanks, weight belts, buoyancy compensators

(BCs) and regulators were transported to different sites in the Bay by automobile. In depths of 10 to 45 feet, and approximately 100 feet west of the Marine Bridge in Jamaica Bay (Fig. 2), we first observed 30-50 Horseshoe Crabs feeding on carpets of mussel beds.

Figure 2. Jamaica Bay

About 25 more crabs were burrowed in the sand seemingly awaiting high tide, to emerge onto beaches to spawn.   I first catch sight of a pair of Horseshoe Crabs swimming effortlessly in front of me, seemingly unaffected by our presence or the low visibility (it is only 2 feet visibility today). It is a typical pair; a smaller male attached at the lower carapace to a larger female (Fig. 3). This is an arrangement that gives the attached male the advantage of fertilizing the eggs first, which is an evolutionary advantage over other nearby males, although many males called satellite males may also try to attach to the same female with varying success (Fig. 4) (Brockmann *et al.*, 1994).

Figure 3. A pair of horseshoe crabs.

On our second dive on June 1, we wanted to see if the crabs were still entering the bay, or if most had already entered. By boat, approximately 1 mile west of the Marine Parkway Bridge, we dove near the entrance of Jamaica Bay from the Harbor Estuary. The currents here are strong because they mix ocean, bay and river currents, and because of this, we did a drift dive rather than anchoring the boat. By letting the boat follow the divers as the current moves them through the water, drift diving requires less energy to swim against the current. No crabs were spotted at this site during this

Figure 4. A satellite male trying to attach to the same female as another male horseshoe crab.

dive, so we assumed that the majority had already entered the Bay to begin their breeding cycle. On our third dive, which began at low tide in the Bay, we observe 100 or more crabs lined up on the sandy incline that led to deeper water, 35 feet out from shore. We did not see any crabs on the flat bottom, which was 40 to 50 feet out into the middle of the Bay. A possible explanation for this is that they are waiting for high tide to emerge on the beach. Of the 40 females we observed that day, all had at least one male attached to their carapace; the rest were males swimming alone. Never having seen any footage of their behavior, it was interesting to observe them in their natural habitat.

Our next dive was done at high tide on June 3, about 40 feet offshore from Breezy Point. We observed approximately 25 pairs of crabs resting in the sand. The occurrence of high tide brings them from approximately 40 feet of water in the middle of the Bay, into the shallower parts near the shoreline. Unfortunately at this site, there was not enough beach left for them to come out of the water and spawn; at high tide, the beach is completely covered by water. Our dive the following week was done at the head of the Cross Bay Bridge, where approximately 100 adults and 50-75 older juvenile crabs (ages 5-8) were burrowed in the sand. Again, most females have a male attached to their carapace. In the middle of the channel near the Marine Parkway Bridge, we observed hundreds of crabs on the sandy bottom a few hours before high tide; most were paired. We noticed similar behavior as in our previous dive, that the crabs are not feeding, and the female was burrowed in the sand. I observed some juveniles (roughly ages 9-10), but most appeared to be adults. Juveniles can be easily recognized because their carapace is a light green compared to a dark brown or black, and is free from scratches frequently seen in older crabs.

The next two dives took place in early August, and we observed hundreds of crabs in the region of the Marine Parkway Bridge, typically swimming paired. Approximately 50 crabs were entangled in fishing line from fishermen casting from the bridge. This is commonly seen because organisms such as mussels encrust on their carapace, which affords a place for line to get caught. Since they would soon die due to malnutrition, we cut away the fishing line to free them. We lifted a few at a time into the boat to be tagged. Our tagging process is simple; a cordless drill is used to make a 3/18' hole in the outer edge of the carapace and a plastic screw attached to a tag is fitted. If someone sees the tag on a Horseshoe Crab, they phone a toll-free 800 number, and it is documented. As of May 1999, two Horseshoe Crabs we tagged last year in Jamaica Bay were spotted the year after in the Bay. Also, tagged Crabs in Staten Island, NY, were seen the nest season in Sandy Hook Bay, Ft. Monmouth, and Manasquan NJ; some were also found back in Jamaica Bay (Table 1).

On our eighth dive, we wanted to see if any crabs remained in the inner part of the Bay in September. In 5 feet of water at the Cross Bay Bridge site, we covered hundreds of feet, seeing few adults but numerous (100-200) juvenile crabs. They appeared to range in age from 1 to 3 years old. Most remained burrowed, their carapace almost completely covered in the sand of the cove bottom. It has been noted that the juveniles are too fragile to make the trip out of the Bay, so they remain here for 7-8 years, molting and growing 25 percent larger with each molt. After 16 molts they are usually between 9 and 11 years of age, and their shells are harder (Shuster, 1958). The shells on the crabs we observed were paper thin and white or light tan in color, but all with the same physical characteristics as the adults. Their size was on average 80-90 mm at the largest part of their prosoma.

Our ninth dive for the season was done near the open harbor waters, where the Bay ends and the NY Bight begins. The reason for diving here at this time, was to see if the crabs were on their way out of the Bay, or if they had already left for the continental shelf. In an extremely strong current and at a very high tide (high water approximately +5.3 feet), we moored the boat at the mouth of the bay. This dive was in about 18 feet of water, in strong

Table 1.

Limulus Tag Data Log Sheet

Limuli Labs
(609) 465-6552
7 Bay Ave.　Dias Creek Cape May Court House, NJ 08210

TAG #	LOCATION TAGGED	DATE TAGGED	AGE	SEX	PROSOMAL WIDTH (mm)	OBSERVATIONS	DATE RECAPTURED	LOCATION RECAPTURED
1	3026	JBWR	06/11/1998	Middle age	Female	220		10/07/1998	Jamaica Bay, NY
2	3042	JBWR	06/11/1998	Young	Female	239	Tag fell off	02/01/2000	Long Beach Island, NJ
3	3055	Staten Island	06/15/1998	Middle age	Female	236	Half tail	06/24/1999	Jamaica Bay, NY
4	3062	Staten Island	06/16/1998	Young	Male	189		12/17/1998	Sandy Hook, NJ
5	3065	Staten Island	06/16/1998	Young	Female	229		05/07/2000	Sea Girt, NJ
6	3066	Staten Island	06/16/1998	Young	Male	162		08/13/1999	Jamaica Bay, NY
7	3073	Staten Island	06/16/1998	Young	Male	186		05/03/1999	Sandy Hook Bay, NJ
8	3075	Staten Island	06/16/1998	Middle age	Male	170		06/19/1998	Great Kills, NY
9	3100	Staten Island	06/16/1998	Middle age	Male	186		08/05/1999	Jones Inlet, NJ
10	3107	Staten Island	06/16/1998	Middle age	Male	102		06/17/1999	Great Kills, NY
11	3135	Staten Island	06/18/1998	Middle age	Male	181		11/26/1999	Bedford, NJ
12	3140	Staten Island	06/18/1998	Middle age	Male	102		03/24/1999	Fort Monmouth, NJ
13	3185	Cross Bay	06/12/1998	Middle age	Female	148		06/20/1998	Jamaica Bay, NY
14	3216	Breezy Point	06/17/1998	Middle age	Female	225		06/20/1998	Crookes Point, NY
15	3217	Breezy Point	06/17/1998	Middle age	Female	191		09/10/2000	Staten Island, NY
16	3218	Breezy Point	06/17/1998	Middle age	Male	170		06/07/2000	Jersey City, NJ
17	3268	Staten Island	06/24/1998	Middle age	Female	225	Forked tail	06/24/1998	Crookes Point, NY
18	3270	Staten Island	06/24/1998	Young	Male	152		06/24/1998	Crookes Point, NY
19	3273	Staten Island	06/24/1998	Young	Male	170		06/26/1998	Crookes Point, NY
20	3284	Staten Island	06/24/1998	Middle age	Male	148		06/26/1998	Great Kills, NY
21	3311	Staten Island	06/30/1998	Old	Male	162		07/03/1998	Great Kills, NY
22	3318	Staten Island	06/30/1998	Middle age	Male	162		04/07/2000	Sandy Hook, NJ
23	3342	Staten Island	07/01/1998	Middle age	Male	155		08/01/1999	Sea Bright, NJ

rip currents. Here we observed hundreds of crabs heading out of the Bay and also being pushed upside down while fighting the current. Some seemed able to feed on mussels as they moved along, but most we observed were single and moving quickly. The last dive we did for the season was on October 22. The water was 65° F, and we found only one or two single crabs swimming in the Bay by themselves. It is now apparent that the majority of crabs left for the deeper waters of the ocean, only to return again next May.

4. DISCUSSION

Having recorded approximately one hour of videotape after editing, it was reviewed by Dr. Carl Schuster, an expert in the field with some 40 years of research and experience. Some of the highlights of the tape included behavior of Horseshoe Crabs while feeding on beds of mussels, swimming in the intertidal zone, and burying in the sand. This last behavior is most likely done to enable them to feed on benthic invertebrates, rather than for protection. In the open ocean, they have been known to be predated upon by sharks, however, man continues to be the primary destructive force on their numbers.

5. CONCLUSION

Most Horseshoe Crabs swim from their deep wintering grounds into estuaries in late March. It has been theorized that, since they all travel at approximately the same time, the stimulus that facilitates this synchrony is the position of the sun. On any given year in March, the temperature of the water and the temperature of the air may be very different from previous years, but what is unchangeable is the arc that the sun makes each day throughout the year.

It is well known that Horseshoe Crab spawning behavior is driven mostly by moon phases and tidal positions (Shuster and Botton, 1985). Horseshoe Crabs emerge together for the first time each year when the moon is in its full phase, in late March, and when the tide is high. The highest tides are usually during the evening, when most emerge to spawn, however many spawn on beaches during the day also. It is apparent that in September the majority of Horseshoe Crabs are not present in Jamaica Bay and have departed for the continental shelf.

Some of the behavior and information we still need to videotape are:
• Juveniles swimming in the intertidal zone to see how this mechanism of locomotion assists them.

- Adults swimming along the bottom of the Bay and whether they use their legs to push off the bottom to assist them in this type of locomotion.
- Adults feeding, and whether both males and females feed when they are attached as a pair.
- Either video or radio tracking of the crabs to their wintering destinations.

REFERENCES

Barlow, R. B. Jr., M. K. Powers, H. Howard, L. Kass, 1986. Migration of *Limulus* for Mating: Relation to Lunar Phase, Tide Height and Sunlight. Biol. Bull. **171**:310-329. (October 1986).

Borowsky, B., P.A. Ander, and J.T. Tanacredi. 1993. The Effects of Low Doses of Waste Crankcase Oil on *Melita nitida* Smith (Crustacea:Amphipoda). J. Exp. Mar. Biol. Ecol., **166**:39-46.

Botton, M.L., and J.W. Ropes. 1987. Populations of Horseshoe Crabs, *Limulus polyphemus*, on a Sandy Beach Biological Bulletin. **173**:289-298.

Brockmann, H.J., T. Colson and W. Potts. 1994. Sperm Competition in Horseshoe Crabs (*Limulus polyphemus*). Beh. Ecol. Sociobio., **35(3)**:153-160.

Cohen, E. 1978. Biomedical Applications of the Horseshoe Crab (Limulidae). Alan R. Liss, Inc. New York, New York.

Itow, T., R.E. Loveland and M.L. Botton. 1998. Developmental Abnormalities in Horseshoe Crab Embryos Caused by Exposure to Heavy Metals. Arch. Environ. Contam. Toxicol, **35**:33-40.

Lenihan, D.J. and J. Brooks. 1998. Underwater Wonders of the National Parks. Compass America Guides, 5332 College Ave., Oakland, CA.

Loveland, R. E., M. L. Botton and C. N. Shuster Jr., 1996. Life History of the American Horseshoe Crab (*Limulus polyphemus* L.) in Delaware Bay and Its Importance as a Commercial Resource.

Shuster, C. N., Jr., 1958. On morphometric and Serological Relationships Within the Limulidae, with Particular Reference to *Limulus polyphemus* (L.). Diss. Abstr. **8**:371-372

Shuster, C.N., Jr. and M.L. Botton. 1985. A Contribution to the Population Biology of Horseshoe Crabs, *Limulus polyphemus* (L.) in Delaware Bay. Estuaries **8(4)**:363-372.

Sullivan, G. 1964 The Complete Book of Skin and Scuba Diving, Longmans Canada Limited, Toronto, pp. 28- 29.

Tanacredi, J.T. and J. Loret (Ed.) 1998. Ocean Pulse a Critical Diagnosis. Plenum Press, 233 Spring Street, New York. pp.56-59.

University of Delaware, Hugh R. Sharp Campus. February 23, 1996. Proceedings of the Horseshoe Crab Forum, Status of the Resource. 60pp.

Chapter 11

Horseshoe Crabs: An Ancient Wonder of New York and a Great Topic for Environmental Education

DON RIEPE
Senior Naturalist, Jamaica Bay Wildlife Refuge, National Park Service

1.   INTRODUCTION

Each spring, between mid-May and mid-June, thousands of fierce-looking creatures crawl from the sea to mate and lay eggs along the sandy shorelines and mudflats of New York City. These harmless animals called horseshoe crabs are not really crabs at all, but more closely related to arachnids (spiders and scorpions). A living fossil, the horseshoe crab evolved long before the dinosaurs with an ancestral heritage dating back to the Triassic Period two hundred million years ago. Currently, four species exist worldwide. One species populates the Atlantic coast from Maine to Mexico with the largest concentrations found in Delaware Bay. Named *Limulus polyphemus* after the one-eyed giant of Greek mythology, this horseshoe crab actually has nine eyes: one large compound eye on each side of its shell, two small ones in the front center and five light-receptive organs underneath. The other three species of horseshoe crabs are found in the Indian and Pacific Oceans. During high tides, especially at new and full moons, these "crabs" emerge from the water to spawn. The larger females are usually accompanied by one or more smaller males that attach themselves to her back by specially adapted clasper claws. At the high tide line, the female will dig a nest in the wet sand and lay tiny greenish eggs. The attached male fertilizes the eggs as they are laid and then both move back to deeper waters to feed on benthic animals such as marine worms, crustaceans, and mollusks.

Limulus in the Limelight, Edited by John T. Tanacredi
Kluwer Academic/Plenum Publishers, New York, 2001
131

About a month later the eggs hatch out, each one containing a tiny, but tail-less replica of the adult horseshoe crab. The little horseshoe crabs will molt their shells several times yearly during the first few years of life and then once yearly thereafter. They reach adult size in about 10-13 years and may live another 7-10 years. Before molting takes place, a new skin forms under the existing shell. The old shell splits open along the front and the crab walks out. The animal then takes in water and digs into the sand. This new skin is stretched larger and hardens around the horseshoe crab to form a new shell. The molted shells can be found along beaches at any time of year.

2.   A SIGHT FOR YOUNG EYES

The eggs provide a bonanza for migrating shorebirds arriving in New York City from their winter homes in Central and South America. Some birds such as Black–bellied Plovers and Red Knots may have traveled several thousand miles across the ocean, making their first landfall in the estuaries of New York and New Jersey. Peak shorebird migration coincides with the peak horseshoe crab egg laying providing critical nourishment for many shorebird species as they head to their Arctic breeding grounds. At the Jamaica Bay Wildlife Refuge in Broad Channel, Queens, I have observed these mating rituals and feeding frenzies for many years and am still amazed by the abundance and diversity of participants.   Besides red knots, sanderlings, ruddy turnstones, and about 20 other species of shorebirds, the egg feast attracts many laughing gulls, glossy ibis and even Canadian geese, whose goslings feed on animal matter as well as vegetation.   During this peak horseshoe crab time in the Park, school children walk the trails, touch these animals in the wild and bring home a special message: conservation of a 350 million year old species.

Snowy egrets join in, not to feed on eggs, but on the mummichogs, silversides, and other small fish that are taking part in the bountiful melee. Larger predators, such as herring and black-backed gulls, will frequently take advantage of overturned horseshoe crabs and peck out their gills, leaving a beachfront strewn with dead and dying horseshoe crabs. Despite this heavy onslaught, the horseshoe crabs keep coming ashore, determined to carry out the reproductive urge as they have done for millions of years before the advent of *Homo sapiens*.

3.   CONCERNS

It is humans, however, that poses the greatest threat. In earlier times, native Americans used the horseshoe crab for food and the shell for bailing water out of their canoes. They also used the long pointed tail, or telson, for spearing fish. None of these uses threatened the horseshoe crab's existence. Today, however, using more efficient trawling techniques, fishermen harvest great numbers of horseshoe crabs for bait and many biologists are seriously concerned for their future.

Shoreline development is another problem. As sea level rises and people continue their migration to live in coastal areas, available shoreline habitats are becoming changed by bulkheading and dredging. Even though most coastal states have laws protecting these wetland areas, there is a continued "nibbling away" of shoreline habitat from use of legal "loopholes" and variances as well as degradation of habitat from other disturbances including offshore dredging and water pollution from increased boating. Sewage outflows and untreated runoff further exacerbate the situation. As our coastal population swells, there will be increased pressure to build groins, seawalls, and other shoreline stabilizing methods used to protect coastal property – all of which impact natural shoreline habitats.

Another human-related problem is the tide of floatable debris littering many shorelines, thus impeding the crabs' access or entangling the animal with monofilament or other plastic.

Fortunately, there is a growing public awareness about this issue. Many volunteer groups routinely clean beaches (i.e. American Littoral Society) in all coastal states, and cities are seeking ways to keep trash from entering waterways. In New York City, the Department of Environmental Protection has purchased several skimmer boats that remove debris from the water's surface.

Often overlooked in the equation is that the horseshoe crab has great medical value to humans. The large compound eye and accessible optic nerve have been used in scientific research for over 50 years. The *Limulus* lateral eye is one of the most thoroughly understood of all sensory receptor systems today. The copper-based blood contains a clotting factor that can detect minute amounts of pathogens. At Woods Hole Oceanographic Institution and other research centers, horseshoe crabs are routinely bled and then released generally unharmed (less than 10% mortality) back into the water. Unlike the red color of human blood, horseshoe crab blood turns bluish when exposed to air. This color comes from hemocyanin, a copper-based molecule that carries oxygen through the circulatory system. An extract of blood cells from the horseshoe crab is used to detect the presence of endotoxins in human blood serum. Chitin, the substance that makes up the horseshoe crab's shell, is used in surgical sutures and bandages that promote healing in humans. One can only imagine what other beneficial secrets are yet to be discovered from studying this living fossil.

4.   CONCLUSION

There is still much to be done through education since these crabs still suffer from a maligned superstition passed down through generations. Horseshoe crab programs are becoming more popular and each spring both Urban and National Park Rangers, as well as school groups, that visit New York City's 26 miles of beaches witness the fascinating story of this prehistoric wonder, the ageless horseshoe crab... a true environmental education.

Chapter 12

Living on *Limulus*

DAVE GRANT
American Littoral Society, Box 539, Sandy Hook, NJ 07732, USA

1. INTRODUCTION

The horseshoe crab or "soldier" crab as it is sometimes called, is arguably the most interesting creature on our coast. Although most people would not put it high on their list of graceful and beautiful animals, it generally leaves a lasting first impression on those who encounter it.

Spaniards exploring Florida's West Coast were impressed; naming Cockroach Bay after a creature that is neither an insect nor a true crab, but more closely related to spiders and scorpions. Understandably, those who were first to note *Limulus polyphemus* probably had little interest in taxonomy, but were more concerned with the practical value of their discoveries.

The French explorers were also impressed by the "king crab", and it is worth noting that Samuel de Champlain's map from his 1604 voyage to "New France" has on its margin, sketches of a few New World creatures that obviously impressed him. One of those animals is a horseshoe crab with the intriguing, presumably Native American word "Sijuenoc" scratched in next to it. The map was widely distributed in its day and I've seen a copy on display at Acadia National Park in Maine.

Much has been written on the horseshoe crab's remarkable life history. In fact, it has been called the most intensively studied marine invertebrate in the world. However, little mention is made of the great variety of creatures that benefit from the crab or live in association with this common denizen of the shallow waters off our shore.

Limulus in the Limelight, Edited by John T. Tanacredi
Kluwer Academic/Plenum Publishers, New York, 2001

135

Over the years I have been compiling an ever-growing list of creatures that somehow rely on horseshoe crabs, and have decided that if I had only one animal to choose as a teaching tool about life in the sea, this venerable arthropod would be on the top of my list. Examining a mature crab is like perusing a text book on invertebrate zoology, and almost any specimen you might pickup has at least two or three other species in tow.

Most of the myriad creatures that are found clinging or growing attached to the horseshoe crab are probably opportunists, but a few residents of the moving menagerie depend on the crab for survival. Looking over a well inhabited crab, I am often reminded of Jonathan Swift's jingle:

Big fleas have little fleas
Upon their backs to bite 'em
And little fleas have lesser fleas
And so, ad infinitum.

2.   LIFE HISTORY

In an evolutionary sense, the horseshoe crab is a conservative fellow, changing little since it first left trails between Paleozoic tidepools hundreds of millions of years ago, enduring great changes through the earth's history. Marine biologist Bill Hall of the University of Delaware likes to say, "when somebody drops the bomb, two things will survive: cockroaches and horseshoe crabs." It is about as close as nature gets to a permanent fixture in this dynamic environment, and like any firm substrate in the sea, its surface is quickly covered by "fouling" organisms that require a safe haven from siltation, as well as access to the water currents that deliver food and oxygen. Because it is large, long-lived and mobile, the horseshoe crab is a magnet for a variety of the invertebrates in the sea as it migrates from the continental shelf to our estuaries to spawn each spring.

A horseshoe crab that has recently shed, especially a rapidly growing youngster, has a smooth and beautiful olive-colored shell that is free of scratches and marine growth. From observations in the aquarium and field, it appears that one way juveniles keep their shells squeaky clean, intentionally or incidentally, is by regularly plowing through the sand and spending extended periods of time completely buried while at rest.

For the aquarist worried about rotting fish in their tank, this disappearing act, which might last for over a week, makes them unnerving pets; but it is all part of the crabs behavioral repertoire in the wild. While diving during rivers the summer, I have observed them burrowing into the sediments to avoid being pushed around by tidal currents. While treading up clams from the marsh in winter, I've dug up young crabs in the sandy creeks, apparently hibernating. This may be where juvenile horseshoe crabs "mysteriously" disappear to during their first few winters, surviving the cold and keeping free of fouling organisms at the same time.

However, the most important feature of their lives that keeps the shell clean is rapid growth, and they shed most frequently when they are juveniles – up to five times a year. This, more than any other factor prevents larvae of other creatures continuously settling out of the plankton community from becoming permanently established on their shells.

For horseshoe crabs, it is, quite literally, a drag getting old. As a crab ages and its growth rate slows, it sheds less frequently and begins to display a striking variety of hitch-hikers. The assortment, size and growth rate of this zoological "5 o'clock shadow", gives us an idea of how recently the crab has shed and comes a gauge of its growth rate; information that is otherwise difficult to obtain from animals that lose their entire shell, and any markers that are placed on them as they grow.

A number of these creatures associate with horseshoe crabs because they are a source of food. Others are permanent residents, apparently living intimately with the horseshoe and nowhere else. Others may be attracted to settle on the crab because of the presence of members of their own species. Most of the other hitch-hikers may settle out of their plankton stage randomly, and with luck, end up living on *Limulus*.

Many animals use the horseshoe crab for food, although the adult crab is so large that few things bother it. The shell of an adult crab is often riddled with scars from its few enemies. People are the worst culprits when the crab is inshore, poking and stabbing them, often because they look monstrous in their faunal overcoats. Some specimens have evenly spaced, but healed gashes across the shell, testimony to their great recuperative capabilities. These are often propeller scars, however, I like to embellish things a bit for curious youngsters and add, "Although it could be the bite of a shark or loggerhead turtle!" perhaps the only two creatures that regularly try to tackle an adult crab.

Even before birth and in death, the horseshoe crab is exploited. Shorebirds are noted for their dependence on the crab's eggs, but I've seen many other birds, even ducks, treading the sand for them at the water's edge. At the high tide mark, I've found tiny nematode worms wriggling among the egg clumps in nests, presumably feeding on the eggs. Gulls pounce on overturned crabs and tear them apart to eat the gill and muscle tissue, and flies and sand fleas are soon attracted to the dead crab and utilize what remains of its flesh. The crab exacts some revenge though; it harbors an encysted flatworm in its gut, which matures into a parasite in the gull's intestine.

As on most other submerged surfaces in the sea, there develops a film of bacteria feeding on organic materials adhering to the crab shell. These pioneers set the stage for the rest of the fouling community of invertebrates, and may ultimately help cause the demise of individual crabs by damaging the shell. On adult crabs in their terminal molt, the microorganisms begin to take a toll, and as time goes on their shell becomes darker and shows more pock-marking from the presence of bacteria that utilize the chitin. It's fortunate there's a bacteria to eat just about anything in this world. Otherwise, after millions of generations of crabs, their shed shells would outnumber even old National Geographic Magazines, and the world would have long ago been buried in them.

3.   HUMAN INFLUENCES

Of all the creatures that utilize the horseshoe crab, only humans seem to harass it from cradle to coffin. Friends in New York City tell me that in Chinatown the spring, the gravid female crabs are sold for their eggs. (It seems that, like every unusual oriental dish, they are reputed to be an aphrodisiac). The tail and leg muscles are also edible and it is said that the horseshoe crabs were eaten by Native Americans.

Even today, around Memorial Day when the crabs are spawning, you can visit Sandy Hook, a popular retreat for recent immigrants living in the New York City area, and occasionally come across a family filling their car trunk with crabs. You can be certain they're not taking the crabs back to the city to be pets.

Horseshoe crabs are also used as bait. Fishermen freeze the females, saw them in half, and insert them into eel traps. The crabs are famous for the fertilizer and farm feed industry they once supported along Delaware Bay, but even up on Raritan Bay "old-timers" tell me that as late as 1950 farmers from Freehold gathered them by the wagonload to plant with melon seeds.

Even the horseshoe crab's remains benefit people. The chitin has been used medicinally, and the complete shell is a popular curio at shell shops outside of the animal's range. I've even come across a design for a Victorian-era lady's purse made from the shell.

The shells can be used for more than whimsical folk art though. Like Native Americans, I've used an empty shell to bail out my canoe and if I ever had the need to fashion a fish spear, my first experiment would be with a horseshoe crab tail.

4. HITCH-HIKERS GALORE

So much for butchering horseshoe crabs or using their remains to exploit them. What I really find interesting are those creatures that are dwelling with crabs (Fig. 1). The most unique confederate is the *Limulus* leech, an inconspicuous, but regular "ectocommensal" found on the underside of the crab. The "leech" is a flatworm, one of the Platyhelminthes; an interesting phylum because most of its members are parasitic – like tapeworms in humans; or commensal – apparently harmless or even beneficial to their host. The *Limulus* leech is not a true "blood-sucking" parasite in its behavior or classification, although you'd think so, judging from the name – *Bdelloura* – that taxonomist borrowed from the Greeks. (A hint to the word's pronunciation and origins is the Bdellometer – pronounced "delometer" – a 19th century medical gadget developed as a substitute for leeches in the Byzantine practice of "bleeding" to cure diseases).

Bdelloura is found around the book gills and leg joints of crabs, especially on older females that have not shed for a long time. For those of us who are squeamish about wiggly things in our hands, they are often present in disconcerting abundance, and are said to be toxic if eaten – just in case you're the adventurous type.

We tend to button hole animals as being either aquatic or terrestrial, overlooking the third lifestyle that is so important and prevalent, that of the distant cousin who appears at the door looking for a place to live. Symbiotic relationships can benefit only one partner (Commensal), both (Mutualism), or be detrimental to one (Parasitism). The Limulus leech exhibits features of all three "isms", at least in the literature, and biologist don't all seem to agree on its exact relationship to the crab.

Traditionally the feeling was that the leech was not a parasite and didn't harm the crab, but merely took advantage of the minute bits of organic material that drifted around while the crab was eating or perhaps grazed the film of fouling growth on the shell surface. More recently biologists have started to reconsider *Bdelloura* as a parasite that may weaken the horseshoe crab enough to contribute to its eventual demise.

The leech lays its eggs in the "pages" of the crab's book gills and these are visible as little dark spots. It may also use the cuticle of the gills as a substrate for chemical activity. In time, these actions weaken the gill surface and allow leakage of seawater and bacteria into the crab's body. This may account for the horseshoe crab's hypersensitivity to bacteria in its blood, which makes it of interest to the medical field. Eventually the horseshoe crab begins to suffer from the onslaught.

The largest of the three Bdelloura species is easily seen and measures up to a half inch in length. Like its two smaller and less conspicuous cousins, it

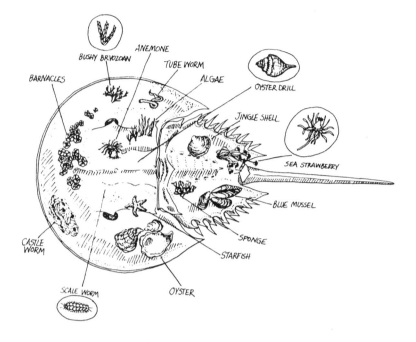

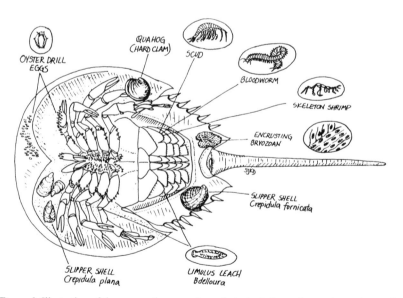

Figure 1. Illustration of the many other organisms that attach themselves to horseshoe crabs. Diagram by Susan Kadolka Draxler.

is known only from the horseshoe crab, but is easy to find if you know where to look. Although the flatworms seem to be more abundant on females, all you need to do is flip over any large horseshoe crab to find a few.

The flatworms appear as small, pale pieces of fleshy tissue that move when you touch them. Under the magnifying glass they are fascinating to watch as they glide across your finger, no doubt trying to escape this hot, alien world onto which they have suddenly been transplanted. How they manage to stay with the crab during the molting process is a bit of a mystery. I've never found them on young crabs and suspect it takes a considerable amount of time or good luck for a community of them to get established. Do they spread between crabs during mating? With the death of their host, how do they deal with such a disaster in their otherwise secured existence?

There are many other more conspicuous residents found on horseshoe crabs. Poring over the phylogenetic tree that forms my ever-growing list, I see nine major phyla of invertebrates regularly represented. Some are probably no better off on a horseshoe crab than they would be on a rock, but others seem to thrive as hitch-hikers and consistently can be found attached at specific spots on the horseshoe crab.

Sponges are filter feeders and are usually restricted to locations where there is enough water movement to bring plankton to them and to prevent burial by siltation. Brightly colored red beard sponges and other fouling *Porifera* occasionally become established on the posterior of horseshoe crabs; probably when the water is cool and the crab is half buried in a dormant stage. They never seem to get a chance to grow very large; at least I've never seen any more than an inch or so long. These slow growers are better suited for cosmopolitan lives on an immovable rock in a current-swept channel where they won't be buried.

Coelenterates are some of the pioneering animals of the fouling community in our waters and are represented on horseshoe crabs by anemones and hydroids. The ghost anemone, a common, non-descript intertidal species, and the colorful striped anemone, an immigrant from Japanese waters, can sometimes be found if the crab's dorsal side is closely inspected. They can only fully be appreciated underwater since they close up when the crab is lifted out of the water. Like the lovely pink sea strawberries, hydroids that are also found on *Limulus*, they must be observed underwater to be fully appreciated. Their exquisite shape is lost when they are not supported by the water.

Snail fur, a stout, bristly cousin of the sea strawberries, is more durable out of the water, but is not as attractive or easy to see without a magnifying glass. However, it is easy to feel and a colony's velcro-like texture contrasts greatly with the smooth glistening shell of the horseshoe crab.

Although the flatworms are well represented by *Bdelloura*, other worms are less common on horseshoe crabs. The *Annelida* or segmented worms are present, but never in such great numbers. Free-living annelids like the scale worm (*Lepidonotus*) or sand and blood worms (*Nereis* and *Glycera*), may be temporary residents, as they would be on a rock or among seaweeds. They may also be potential food items that the horseshoe crab has dug out of the sediments. It is difficult to decide which, but it is not unusual to pull a horseshoe crab out of a net and find a worm or two gliding among its fouling growth community or trapped by the surface tension of the film of water on the crab's shell.

Tube-building worms are common on the backs of horseshoe crabs. Older crabs oftentimes have a filigree of tunnels intertwined on the highest portion of the shell where it is rarely buried. Some worm species, like *Sabellaria*, glue sand grains together to form a protective tube. Obviously the crab must have spent time on sandy bottom for the sand castle worm to collect construction materials, so this helps us trace its movements.

Other worms extract calcium from the water to make their home, and the bottom type where the crab lives is not a factor. The loosely-coiled *Hydroides* is fairly common on our crabs in the Middle-Atlantic, and I've been told the tightly coiled *Spirorbis* worm is found on crabs in the cool waters at the crabs northern limit in Frenchman's Bay, Maine, where Champlain may have encountered them.

Another ancient group, the Echinodermata, is sometimes represented on horseshoe crabs by the starfish. I've never found more than one or two tiny ones on a heavily encrusted crab, so they are probably only temporary residents and fall or move off in a short time. It's hard to imagine they could be any threat to the horseshoe crab since their prey is bivalves, however, in captivity, aquarists have reported adult starfish and urchins grazing on the eyes of resting horseshoe crabs. Perhaps there is also some hazard in the wild for dormant crabs. Of the horseshoe crabs that arrive earliest in the spring at Sandy Hook, a surprising percentage have damaged eyes. Most seem disoriented and make landfall on the wrong side of their destination, ending up stranded by the rough surf on the ocean side and becoming food for gulls.

Several crustaceans are regular companions of the horseshoe crab. Mud crabs and sand shrimp are usually with the horseshoe crabs that we drag up with nets. The shrimp are no doubt incidentals in the catch and temporarily caught up with the crabs legs, although while diving and watching horseshoe crabs, I have seen shrimp and mud crabs hiding on, under, and around their shells as though the resting horseshoe crabs were algae-covered rocks. Tiny juvenile spider and rock crabs also find a secure, prefabricated home in the crevices inside *Limulus* molts.

Give a crab a good shake in a bucket of clean water and after you remove it you'll be surprised at what's swimming around. Scuds (*amphipods*) and marine sowbugs (*isopods*) are likely to be found with horseshoe crabs, especially if they are heavily covered with growth. Both animals are quite common where the crabs pass through seaweed beds, and although they are strong and graceful swimmers, they are also good fish food, so it is certainly to their advantage to stick close to seaweed, rocks, or horseshoe crabs that look like seaweed-covered rocks.

Some years the bottom of the bay is carpeted with skeleton "shrimp" (*amphipods*), and they too can be found on the fouling growth of the crab. The inch-worm-like movements of these tiny crustaceans is often overlooked, especially when the substrate they are on is out of the water, but they are easy to spot if the horseshoe crab is submerged.

There is no mistaking barnacles, and they regularly settle and attach permanently to the backs of horseshoe crabs. In fact they are the crustaceans you are most likely to find on *Limulus*. The crown they form on the cephalothorx is a mark of how deep the crab burrows into the substrate when it is at rest for long periods. Usually the circular scars of detached barnacles are also present, evidence that the crab has not shed for some time and burrowed deep enough at some point to smother the previous generation of barnacles.

The *Mollusca* are better represented on the horseshoe crab than any other phylum. Several species of bivalves become attached to the crabs, and a number of snails are also regularly found gliding around on them.

Oyster spat need to settle on a firm, silt-free substrate, and crab backs occasionally fit the bill, especially in Delaware Bay farther south. Like barnacles, if they die after being smothered in the mud by the crab's burrowing habits, they leave behind scars on the lower shell.

Mussels also like to call the crab home and usually attach themselves near the hinge where water is circulated to the gills by the resting crab. Less frequently, they can be found on the underside of the crab around the legs and gills, and I've freed quite a few old souls whose movements were greatly impeded by clumps of sizable mussels.

The water of Delaware Bay is sometimes too warm in the summer for edible blue mussels to thrive and grow large, but traditionally, residents in Fortesque (NJ) are said to have collected suitable mussels from crabs that move into the bay after wintering offshore. Up in Raritan Bay, we also find horse mussels on crabs in the spring as they arrive to spawn and small ribbed mussels on them after they've been up in the marshes during the summer.

Turn over a horseshoe crab in the shallows and you are likely to discover young soft and hard shell clams, and tiny gem shells hung up in the bristles surrounding its mouth. These are prey items of the crab, the dominant predator of the bay's benthic community and the reason why in the unenlightened and not too distant past, Massachusetts had a bounty on *Limulus*.

Sometimes you'll find the prey exacting a penalty too. As the crab treads the bottom for seed clams, it may inadvertently stick a claw into the open shell of a hard clam that is filtering water. A large clam can clamp down so tightly neither creature can free itself. If the waters were cleaner here on Sandy Hook Bay, on a lunch-time stroll in May I could easily collect enough quahogs from those crabs that are dragging around a living ball-and-chain to make myself a nice chowder. And without even getting wet above the ankles.

Asking around in New England, I've heard of one more bivalve that's been reported attached to *Limulus*, the jingle shell. The delicate and colorful jingle shell secures itself with byssal threads that pass through its lower shell. It then grows along the contours of the shell or rock it calls home. With a geographical range that extends south to the Yucatan, there are probably many other bivalves that take up residence on the horseshoe crab and hopefully, I'll eventually hear about them from readers.

Another mollusk that is form fitted to its substrate is the slipper shell. Three species – the common, convex and flat – are regularly found attached to the underside of the crab. They are so abundant and such a regular fixture on *Limulus*, people often mistake them for part of the crab's shell. To insure that there are males and females living in clusters, slipper shells are said the use pheromones to attract larva out of the plankton to settle on their own kind. The horseshoe crab is such a good home to slippers that several generations are usually present in one stack.

Snails are well-represented on horseshoe crabs. I've found periwinkles, basket and mud snails and that old nemesis of the oyster, the drill. The periwinkles graze algae – common on horseshoe crabs that spend time in shallow water. Basket and mud snails are scavengers and swarm to dead horseshoe crabs for food. Snails also lay eggs on the back of the crab, and in the spring we find specimens that are literally carpeted from head to tail with drill and mud snail egg capsules.

Algal growth on the shell is a clue to where the crab has been for the last few weeks. Green seaweeds like sea lettuce and *Enteromorpha*, and brown weeds, like popweed (*Fucus*), thrive best in shallow waters where the light is intense. They also endure desiccation, so we find them getting established on the crab's back in the warmer weather when spawning and feeding in shallow waters. The more delicate red seaweeds, *Agardhiella* and *Gracilaria*, are more likely to be found on crabs in the fall when they are dredged up from the cooler water in the bay.

Bryozoans, the colonial "moss animals", are a difficult group to recognize, but they are often the most abundant hitch-hikers on horseshoe crabs, sometimes forming a coating over most of the crab's shell, even on parts of its underside. The most noticeable bryozoan, *Bugula*, is a bushy, golden seaweed-like creature. It is probably the most attractive ornament found on a horseshoe horseshoe crab. Harvested elsewhere, dried and dyed green, it is familiar to us as the "plant" sold by a florist as "everlasting" Irish sea ferns.

Other bryozoans include calcified types that form a delicate lacy pattern on the top of some shells. A third type, the spongy bryozoans, often spread across most of the upper shell of horseshoe crabs, even covering their eyes.

No one is quite sure how long horseshoe crabs live (although they might live well into a second decade) or on which birthday they reach maturity (although it probably takes about ten years). However, it is suspected that they stop growing at that point, and the spreading growth of the bryozoan colony, especially on adult males, seems to confirm that they indeed stop molting at maturity.

5.   CONCLUSIONS

One of the things I look forward to the most each spring is the horseshoe crabs moving inshore to spawn, because they always bring along something new and interesting to show me. Over the years, I have noticed that the most sluggish are likely to have most of the shell and eyes covered with fouling growth.

Such a thorough covering across the surface of the horseshoe crab may cause it more than a little inconvenience, by interfering with light detection. The thick coating of hitch-hikers seems to presage the final demise of the horseshoe crab. This and their sluggishness are clues that they are suffering and probably won't survive another season.

I pity them and always peel off the growth from the eyes of heavily infested horseshoe crabs, but it probably doesn't help them for long, since they undoubtedly grow right back. In spite of it all, the horseshoe crabs struggle on, ambling off into deeper water, disappearing beneath a thickening blanket of bryozoans, barnacles, and whatever else issues from Triton's wreathed horn. Perhaps, like their terrestrial counterparts, old soldier crabs never die, they just fade away.

Chapter 13

An Estimate of Population Sizes of Two Horseshoe Crab (*Limulus polyphemus*) Sites in Jamaica Bay

DIANA H. HANNA
Brooklyn College, CUNY

1. INTRODUCTION

Although it is referred to as a crab, and has the superficial appearance of a crustacean, the Atlantic Horseshoe Crab (*Limulus polyphemus*) is actually more closely related to arachnids than to Brachyura. The Atlantic species, along with its three Asian relatives, are among the most studied and highly valued invertebrate animals. Since the second half of this century, *L. polyphemus* has been used in various biomedical applications. A protein called limulus amoebocyte lysate, or LAL, which occurs in the blood of horseshoe crabs, is used to screen for contaminants in drugs and medical devices, and is required by the U.S. Food and Drug Administration for testing all drugs used by humans. In addition, the eggs, which result from the mass spring spawning of *L. polyphemus* are an important source of food for several species of migratory shorebird, some endangered or threatened. Semi-palmated Sandpiper (*Calidris pusilla*), Sanderling (*Calidris alba*), Ruddy Turnstone (*Arenaria interpres*) and Red Knot (*Calidris canutus*) are four of the species of shorebird which rely upon the mass spawning of *L. polyphemus* for their nutritional needs. In their northward migration every spring, the birds make very few stops (often only one). It is at this time that they feed upon the eggs of the Horseshoe Crabs, consuming two to three times their weight, before continuing their migration northward to the Arctic Circle (Hall, 1989). If the stock of horseshoe crab eggs should disappear, there may not be sufficient alternative nutritional reserves available.

Limulus in the Limelight, Edited by John T. Tanacredi
Kluwer Academic/Plenum Publishers, New York, 2001

Atlantic Horseshoe Crabs have been documented to leave the sub-littoral zone and approach the shore en masse during the spring high tides in order to take part in the mass spawning (Botton and Loveland, 1987). Normally, *L. polyphemus* oviposit their eggs 10 to 20 cm beneath the surface, too deep for most shorebirds to detect or access. When the spawning crab population reaches a critical density, however, nests buried during previous spawning sessions are dug up in the process of laying newer nests. The eggs from these nests are then accessible to the shorebirds as a nutritional source. This phenomenon has been well documented along the shores of the Delaware Bay, the most popular spring stopover point for northerly migrating shorebirds on the Atlantic coast. The slow moving, easily stranded crabs prefer protected, low-energy beaches in order to disperse laterally along the shoreline and oviposit (Penn and Brockmann, 1994).

There are no previously documented studies of the Horseshoe Crab population in Jamaica Bay. The principle purpose of my survey, therefore, was to quantify the abundance of the population of *L. polyphemus* in the waters close to the Jamaica Bay shore. Based on my sample surveys, I recommend that the population may be large enough to justify an on-going, full-scale investigation.

2.   METHODS

Because the vast majority of the adult life of the horseshoe crab is spent in the sub-littoral zone, it is difficult to get an accurate estimate of the true population numbers. The commonly utilized method of estimating the population size has been to take a count of adults during their annual late-spring, early summer mass spawning. The crabs are known to come out in their greatest numbers at this time and at the spring high tides. These spring tides occur during the full and new moon phases of the lunar cycle. Only natural light was utilized for all of the surveys; therefore observations were limited to the dates in which the high tide occurred before dusk.

I set out to make a baseline estimate of the population in the most time and energy efficient method possible. With this in mind, I limited my survey to two stretches of shoreline. The choice of the two beaches was based primarily on accessibility and the likelihood of finding a robust population of crabs there. The two beaches chosen were:

(1) The northern tip of the Jamaica Bay Wildlife Refuge just north of the Unit 101 site. The sandy beach here is accessible to humans for 126 meters before interruption on the southern end by an outcropping of grassy marsh.

(2) The length of Plum Beach, which measures 1530 meters, slightly under a mile. The beach is bordered to the west by a jetty of rocks and boulders, and to the east by a marsh, which ends the accessible area.

L. polyphemus is quite a slow-moving animal. Due to this and the fact that both of these sites encompass a very short distance, I was able to take a total count of each site during every survey session. This involved counting every live individual on shore as well as all those that could be detected under water. I waded as far into the bay as the winds, turbidity and visibility of the waters allowed for detection of horseshoe crabs. This distance varied from 2 to 16 meters. Because the bay waters were often opaque with sand and other suspended materials, there is a strong likelihood that the horseshoe crabs were undercounted.

At Plum Beach a single count was made, beginning at the extreme western end of the beach moving eastward. For the surveys at the Refuge, two counts were taken. First, the animals were counted moving from the north to the south end, followed by a re-count taken from south to north. The higher number of these two counts for each date is reported.

3.   RESULTS

Table 1 lists the location, time, atmospheric conditions, water temperature and conditions, lunar data and gross population counts for each study date. All observation dates were in 1998.

Table 1. Survey dates, conditions and gross numbers of *L. polyphemus* observed.

Date and time	Location	air temp/ conditions	water temp/ conditions	lunar data	# of animals
5/25 8:45 AM	Plum	Ensuing thunderstorm	---	new moon	18*
6/6 6:00 PM	Refuge	18°C, breezy	---	full minus 3 days	76
6/7 6:30 PM	Plum	winds very high, cloudy	Extremely rough tides	full minus 2 days	83
6/8 7:20 PM	Refuge	clear, 21°C	17°C	full minus 1 day	62
6/9 7:00 PM	Plum	upper 20.6°C, cloudy	20 to 24°C	full moon	1,272
6/10 8:15 PM	Refuge	21 to 17.8°C	26.5°C	full plus 1 day	126
6/26 8:35 PM	Plum	29.4°C	23.5°C	new plus 2 days	439

*This date's survey was cut short due to a thunderstorm.

The ratio of males to females observed during the mass-spawning phenomenon in other areas is invariably highly biased in favor of males. The observations of *L. polyphemus* in Jamaica Bay are no exception. Though the observed male to female ratio varied widely within each study site, a mean ratio of males to females was 3.6 and 2.6 at Plum Beach and the Wildlife Refuge, respectively (Table 2).

Many animals were observed in "pairs" (i.e. pre-copulatory amplexus), and notation was made as to whether the pairs were engaged in moving or nesting behavior. The term pair, here, is defined as any observation of two animals in amplexus. The assumption was made that the animal in front is female. In a great number of cases one female was observed with as many as six to eight males in tandem. In these instances the group was counted as one pair, plus the number of additional males as single. These assumptions are applied when calculating the male to female total ratio. The findings are illustrated in Table 2.

Table 2. Data on behavior categories and male/female ratio observed at each site

Date, time and location	single males	single females	moving pairs	nesting pairs	m/f total ratio
5/25 8:45 AM Plum	---	---	---	---	----
6/7 6:30 PM Plum	23	4	9	19	1.6:1 (51:32)
6/9 7:00 PM Plum	795	5	79	157	4.3:1 (1031:241)
6/26 8:35 PM Plum	290	1	7	67	4.9:1 (364:75)
Plum Beach average	369	3	32	81	3.6:1
6/6 6:00 PM refuge	24	0	12	14	1.9:1 (50:26)
6/8 7:20 PM refuge	31	11	5	5	2.0:1 (41:21)
6/10 8:15 PM refuge	77	1	14	10	4.0:1 (101:25)
Refuge average	44	4	10	10	2.6:1

On two of the three occasions in which a count was concluded at Plum Beach, a significantly large number of adult crabs were observed spawning at high tide. A simple calculation of the density of *L. polyphemus* could be made, based on gross numbers of animals per meter of shoreline. Based on the highest population count, this comes out to 0.83 individuals per meter of shoreline. There appears to be a more random, or less clumped, distribution at the Wildlife Refuge location. The gross population estimate for this 126 meter stretch of shoreline, based, again, on the highest count, turns out to be one animal per meter of shoreline.

4. DISCUSSION

The count taken on June 9th at Plum Beach provides an indication of a viable *L. polyphemus* population in Jamaica Bay. The observation of almost 1,272 spawning individuals on that evening is three times as high as the next highest count at that location of 439 individuals on June 26. Several obvious reasons for this discrepancy can be seen. The full moon occurred on June 9, providing the highest tide except for the new moon. In addition, the time of the high tide on June 9 was 1.5 hours earlier than on June 26. Therefore more daylight was available on this date for observing the crabs. Finally, the end of June may be past the peak of the spawning season.

Because the distribution of horseshoe crabs on Plum Beach is neither random nor uniform, I would not recommend that the above results be used as a population estimate for either Jamaica Bay or the Plumb Beach region. On each survey session *L. polyphemus* was found in greatest numbers beginning at roughly 600 meters east of the benches at the parking lot. On every occasion the vast majority of individuals were found from this point, increasing in frequency up until the eastern-most tip of the beach. In addition, a few individuals were observed on every occasion at the extreme western tip of the beach, amongst the rocky substrate of the natural jetty. If the density were examined for only the stretches in which crabs were found in abundance, the population estimates would be much greater.

Contrary to initial speculation based on raw numbers at each site, the gross density (individuals per linear meter of shoreline) at the Wildlife Refuge is somewhat higher than at Plum Beach (1.0 and 0.83, respectively). The Refuge, however, is in closer proximity to human usage than the Plum Beach location, which is separated from its parking lot by a marshy bank. This site is in juxtaposition with the Wildlife Refuge and may, therefore, also be considered protected. Alternatively, Plum Beach is bordered along its northern perimeter by a major, six-lane highway. These facts do not support or disprove a human proximity hypothesis, but indicate that further investigation of the variable ecology of the shore should be made before a hypothesis is proposed.

5. ADDITIONAL CONSIDERATIONS

Along which other shores of Jamaica Bay are the crabs spawning in significant numbers? What is the magnitude of the populations in the most crab dense regions? Do the areas with the highest density of crabs attract and/or support a significant number of migratory shorebirds? Does the population at Jamaica Bay, especially at Plum Beach, provide a significant amount of food matter (via eggs) for the migrating shorebirds? A count, via the core sampling method, should be taken of on-shore nests to estimate the extent of the offspring generation.

In both locations observations of other species were recorded. Several species of gulls and terns were prominent at both sites. Pigeons, doves, starlings, sparrows and crows were also seen at both locations. An unidentified sandpiper species, a Red-winged Blackbird, and sea robins (fish) were observed at Plumb Beach. Additionally, Mallards and Canada Geese were common in the marshland close to the Refuge study site. How do these and other species affect the viability of the *L. polyphemus* population? How do the various commensal and parasitic "hitchhiking" invertebrates affect the health of their hosts, the horseshoe crab?

To what extent does human interference such as fishing and pollution affect the density of horseshoe crabs? One hypothesis for the very clustered distribution found at Plum Beach follows the argument that *L. polyphemus* avoids the areas that are frequently or heavily populated by humans. The rocky outcropping at the western end is somewhat regularly used as a fishing pier. The most likely explanations, however, will be based on the ecological irregularities of the shoreline.

Finally, can direct human interventions, for example re-orienting stranded crabs, or implementing massive beach cleanup regulations, help the crab population thrive?

ACKNOWLEDGMENTS

My sincere appreciation and gratitude to the following people. Dr. David R. Franz, Janice Emord, Dr. Mark Botton, Dr. Bob Loveland, Don Riepe, Denise Annie Way and Dr. Joan Davis. Due to their invaluable knowledge, guidance and assistance, I have been able to conduct this research.

REFERENCES

Botton, M.L., and H.H. Haskin. 1984. Distribution and Feeding of the Horseshoe Crabs, *Limulus polyphemus*, on the Continental Shelf off New Jersey. *Fishery Bulletin.* 82 (2): 383-389.

Botton, M.L., and R.E. Loveland. 1987. Orientation of the Horseshoe Crab, *Limulus polyphemus*, on a sandy beach. *Biological Bulletin.* 173 (October): 289-298.

Botton, M.L., and J.W. Ropes. 1987. Populations of the Horseshoe Crabs, *Limulus polyphemus*, on the Northwestern Atlantic Continental Shelf. *Fishery Bulletin.* 85 (4): 805-812.

Hall, W.R., Jr. 1989. The Horseshoe Crab – A Reminder of Delaware's Past. *MAS Bulletin, University of Delaware Sea Grant College Program.*

Penn, D., and J. Brockmann. 1994. Nest-Site Selection in the Horseshoe Crab, *Limulus polyphemus.* *Biological Bulletin.* 187 (Dec.): 373-384.

Schuster, C.N., Jr., and M.L. Botton. 1985. A Contribution to the Population Biology of Horseshoe Crabs, *Limulus polyphemus* (L.), in Delaware Bay. *Estuaries.* 8 (4): 363-372.

PART V

A Final Word

Photo by Christine Kurtzke.

"Throughout the 4.6 billion years of its existence, our planet has accumulated an impressive inventory of living plants and animals, a diverse gene pool that should be protected and passed on as our most valuable asset to future generations. There is probably no more compelling example of the importance of maintaining the world's gene pool than that of the horseshoe crab. It is the simplicity of its primitive system that has made the horseshoe crab so well adapted to its shallow-water environment. Had horseshoe crabs altered over time or become extinct, mankind would have lost the genes for a unique biological organism which is now saving human lives due to this primitive heritage. With luck, research and a helping hand, horseshoe crabs should be around for a million (more) years. Can the same be said of man?"

William Sargent, in *The Year of the Crab*

Chapter 14

Preserving a Living Fossil

NILES ELDREDGE
Curator, Division of Paleontology – Invertebrates, American Museum of Natural History, Central Park West at 79th Street, New York 10024, USA

It is all too clear, in reading the contributions to this book, that *Limulus polyphemus,* the American horseshoe crab, is under dire threat of extinction. I find this both appalling and amazing. I am appalled because such a valiant old soldier, whose lineage goes back in recognizable form hundreds of millions of years, only now, in our lifetime, may face extinction and at that by our own hand. And I am amazed because horseshoe crabs are true survivors whose forebears sailed through many a mass extinction event in the geological past while dinosaurs, ammonites and hosts of other creatures succumbed. This suggests to me, perhaps more clearly than ever before, that the power of human destructiveness on the planet may be even more encompassing, more devastating, than the enormous shock caused by the collision between Earth and one or more asteroids or comets 65 million years ago.

Horseshoe crabs are true "living fossils"- and this is no empty phrase. Steve Stanley and I edited a volume devoted to the phenomenon of "living fossils" (Eldredge and Stanley, 1984). From all the varied contributions, we discovered some common themes. Most so-called "living fossils" are anatomically conservative, retaining more than a passing resemblance to long-dead ancestors. They are also relatively isolated in a phylogenetic sense: they are usually species (or small groups of species, like the modern horseshoe crabs) with no truly close living relatives. They stand out, in an evolutionary sense, from everything else.

Another thing about living fossils: they are almost invariably "eurytopes"—able to tolerate wide ranges of ecological parameters—such as temperature, food resources and even oxygen availability and salinity. Yet their closest relatives are often very narrowly adapted species. Narrowly

Limulus in the Limelight, Edited by John T. Tanacredi
Kluwer Academic/Plenum Publishers, New York, 2001

adapted species tend to evolve quickly—but they are also rather prone to extinction—having put all their ecological eggs, so to speak, in a relatively narrow basket. Ecological generalists, on the other hand, can swing with the changing times; they may not evolve very quickly, but they also tend to last longer: while no species is extinction-proof, ecological generalists as a rule outlive their more narrow-niched kin.

Limulus polyphemus, as these contributions make abundantly clear, fits this profile very closely. True, we do not know much about collateral kin to horseshoe crabs in the Upper Paleozoic (from the time, say, in the Upper Devonian/Lower Carboniferous when the opisthosoma was finally welded into a single shield-like skeletal structure). I am tempted to suggest that the various Carboniferous species found variously in marine, brackish and fresh water habitats of "Coal Measures" environments constitute the more narrowly adapted (and long-since extinct) specialized species, while modern horseshoe crabs derive from the one surviving lineage represented by *Paleolimulus* (at first glance, at least, a miniature version of today's horseshoe crabs). There's more to say, of course, about what makes a lineage a "living fossil," but it suits our purpose here merely to note that the concept has meaning—and that modern horseshoe crabs fit the description.

So why, after all these millions of years, do we see a very real possibility that this lineage has not long to go? The papers in this volume bring home trenchantly the salient fact that human actions are poised to accomplish what no asteroid or global cooling event of Earth's long past could do: eradicate the modest, yet incredibly old-aged horseshoe crab line. Overfishing (even though most cultures do not eat them—though I saw a specimen of, I believe, *Tachypleus tridentatus,* kept in a tank in a seafood restaurant in China recently), pollution, and perhaps the most subtle and ravaging nemesis of all, habitat destruction—these are the unholy threesome that may well doom this lineage.

As is usual, we always ask, why should we care? Maybe being around for several hundreds of millions of years as an unbroken conservative evolutionary line might make them sentimental favorites; after all, the earth has pretty much always had horseshoe crabs ever since macroscopic life appeared in the oceans over ½ billion years ago! But, as this volume also teaches us, there are more proximate and pressing reasons why we should be alarmed over the impending doom these creatures face. They are useful in biomedical research. They are useful as food and bait (yes, that's right, I am indeed saying that one of the threats to their existence stems from one of the reasons we need them around: after all, it's over-fishing that poses the threat). Intellectually, too, horseshoe crabs have much to tell us about the anatomies and behaviors of long-dead collateral kin—such as my own much beloved trilobites—gone these past 245 million years.

Yet there is more. We often pay lip service to the vital role this or that species plays in its ecosystem—yet sometimes it is difficult to be precise about what damage to the system would be caused if a particular species were all of a sudden to disappear. But with the American horseshoe crab, I think we have one of the clearest examples imaginable—for the horseshoe crabs, especially but not exclusively, in Delaware Bay, are linked to ecosystems in such far-flung places, that it is safe to say their demise would have a global impact. I refer, of course, to the feeding frenzy of Red-Knots, Laughing gulls, Ruddy Turnstones and other species of migratory birds—who stop off to gorge on horseshoe crab eggs while on their way to breeding grounds farther north. Through the year some of these bird species cover vast territories, impinging on tundra, interior wetlands and shoreline habitats. Horseshoe crab eggs in abundance are a vital cog in this migratory wheel.

Imperil the horseshoe crab; imperil the red knot and others who dine on those eggs. Imperil these bird species, and there are sure to be cascading effects on the biotas in each of the disparate ecosystems in which they breed and spend the winter—not to mention the habitats they visit along the way.

But, as I said at the outset, what is perhaps most disturbing to me is that, with the horseshoe crab under such threat, we have one of the clearest signals I have yet encountered that perhaps the destructive hand of humanity might soon equal or even overshadow the explosion that took out the dinosaurs, ammonites and so much else 65 million years ago.

I well remember newspaper accounts from the 1960s or 1970s, recounting pollution in the Raritan River and Bay—and how horseshoe crabs, those tough guys who can stand just about anything, were, as expected, still hanging in there when pretty much everything else was gone. A cleanup of the Raritan system temporarily, at least, restored some of the stocks of those other aquatic species. But the conversion of habitat, production of pollutants and demand for increasing "takes" of horseshoe crabs along with all other commercially exploited species is expanding relentlessly. It threatens to pass the point of no return—even for that doggedly hardy species *Limulus polyphemus*.

There are some good, strong explicit suggestions in this volume directed at the question: what can we, what *should* we, do to save the American horseshoe crab? I am heartened by the specific suggestions for promulgation of regulations governing catch size, coastal development and pollution. We need a tough stance to protect this no-longer tough customer—this North American horseshoe crab—who, after all these eons, may at last have met its match in *Homo sapiens*.

REFERENCES

Eldredge, N. and S.M. Stanley (eds.). (1984) *Living Fossils.* Springer-Verlag, New York and Berlin.

Horseshoe Crab eggs – fertilized. Photo by Don Riepe.

Contributors

Dr. Mark Botton is Professor of Biology and chairman of the Department of Natural Sciences at Fordham College at Lincoln Centre. He has been studying Horseshoe Crabs since 1976 and received his Ph.D. in 1982 from Rutgers University researching *Limulus*.

Josh Eagle is an attorney with the National Audubon Society. He has a MS degree in Forest Sciences from Colorado State University where he focused his research on the intersection of law, economics and ecology.

Dr. Sylvia Earle is a distinguished oceanographer and marine biologist. Dr. Earle is presently chairwoman of Deep Ocean Explorations and Research (DOER) marine operation. She is the National Geographic's Societies Explorer in Residence and the Centre for Marine Conservation's Ambassador for Oceans. Dr. Earle has over 100 publications including the popular Sea Change (1995) and Wild Ocean (1999). She has had numerous awards and has led more than 50 expeditions worldwide spending over 6,000 hours underwater. She has most recently (2000) been appointed to the National Parks Foundation.

Dr. Niles Eldredge has been a paleontologist on the curatorial staff of the American Museum of Natural History since 1969. His speciality is the evolution of trilobites – a group of extinct arthropods that lived between 535 and 245 million years ago. Dr. Eldredge's main professional passion is evolution. Throughout his career, he has used repeated patterns in the history of life to refine ideas on how the evolutionary process actually works. The theory of "punctuated equilibria", developed with Stephen Jay Gould in 1972, was an early milestone. Dr. Eldredge went on to develop a hierarchical vision of evolutionary and ecological systems, and in his book *The Pattern of Evolution* (1999) has recently developed a comprehensive theory (the "sloshing bucket") that specifies in detail how environmental change governs the evolutionary process. Concerned with the rapid destruction of many of the world's habitats and species, Dr. Eldredge was Curator in Chief of the American Museum's Hall of Biodiversity (May, 1998), and has written several books on the subject – most recently (1998) *Life in the Balance*. He has also combated the creationist movement through lectures, articles and books – including *The Triumph of Evolution... And the failure of Creationism* (2000).

David Grant is the American Littoral Society's Chief Naturalist and member of the faculty at Brookdale Community College in New Jersey.

Diana H. Hanna is a graduate student at Brooklyn College City University of New York under Dr. David Franz, Biology Department

Peter Himchak is with the New Jersey Division of Fish and Game as a supervisor Biologist. He is presently supervising the Nacote Creek Research State. He has a BS in Biology from St. Peter's College and an MS in Ecology and Fisheries from Rutgers University.

Christine Kurtzke is the fishery Biologist for the National Park Service at Gateway National Recreation Area. She has an MS degree in Biology from Long Island University, and is a Ph.D candidate at the City University of New York. Her interests are in the reproductive success of species found in urban estuaries.

Dr. Robert Loveland is Professor of Biology at Rutgers University and has researched the biology of Horseshoe crabs and their relationship to migratory birds for over 20 years.

Dr. Thomas J. Novinsky is President and Chief Executive Officer of Associates of Cape Cod, Inc., a world-leading supplier of *Limulus Amebocyte Lysate* (LAL). He is recognised worldwide and as an expert in endotoxin removal and LAL testing in the pharmaceutical industry. He holds 4 patents and has over 60 publications. He has a Ph.D. in microbiology from the University of Kansas and a BS in microbiology from Penn state University. His current research interests include clinical and environmental implications of bacterial endotoxins and fungal glucans and a development of a synthetic replacement for LAL.

Donald Riepe is the Chief Naturalist at the Jamaica Bay Wildlife Refuge at the National Park Service, Gateway National Recreation Area. He has used Horseshoe Crabs as a significant part of the interpretive activities of the Refuge for more than 25 years.

Dr. Carl N. Schuster, Jr. is Adjunct Professor of Marine Science at the Virginia Institute/School of Marine Science, The College of William and Mary. He began his studies of horseshoes crabs over 50 years ago when he worked on the digestive tract of larval stages for his MS from Rutgers University in 1948. He went on to study and research at Woods Hole Oceanographic Institution and received his Ph.D. from New York University on Limulus. He is publishing a treatise on *Limulus* by Harvard University Press.

Benji Lynn Swan works for Limuli Associates, Inc. and is responsible for the harvesting of crabs for pharmaceutical research and medical purposes.

Dr. John T. Tanacredi is Chief, Division of Natural Resources for the National Park Service at Gateway National Recreation Area. He is a Research Associate in the Department of Invertebrate Zoology at the American Museum of Natural History and at the Aquarium for Wildlife Conservation, Wildlife Conservation society. He is an Adjunct and Full Professor of Environmental Sciences at Hofstra University and Polytechnic University. He has over 50 scientific publications and two books: the most recent (1997) *Ocean Pulse: A Critical Diagnosis* by Plenum Press Ltd.

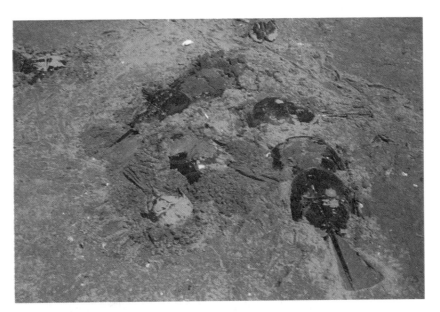

Male Horseshoe Crabs congregate around a female crab in sediment. Photo by John Tanacredi.

Horseshoe Crab trail, Jamaica Bay. Photo by John Tanacredi.

Index

169